Under the Skin, Above the Pavement

Under the Skin, Above the Pavement

Carlin D Nelson

NEXUS

OVERVIEW

Under the Skin, Above the Pavement is a groundbreaking exploration of how the environments that shape men, particularly Black, brown, and marginalized men in urban settings become embodied across systems of biology, behavior, and identity. This book uses an intersectional framework to examine how masculinity, race, class, sexuality, and place converge to shape health outcomes, revealing the deep connections between systemic inequality and the biology of survival. Drawing from public health research, lived experiences, cultural analysis, and interdisciplinary studies, this book explores how chronic stress, environmental degradation, housing instability, food insecurity, and hyper-surveillance impact men biologically. These experiences translate into tangible physiological outcomes, from heightened inflammation and disrupted hormone regulation to changes in gene expression and suppressed immunity. Masculinity is framed not as a singular identity but as a complex risk factor that interacts with these systems and compounds their effects on health.

This book also delves into how men inherit not just genetic material, but cultural legacies of caregiving, trauma, and resilience. By exploring male caregiving, fatherhood, and intergenerational impacts, it shows how these roles are shaped by both biological and social pressures. Each chapter delves into a distinct facet of embodied risk: the physiological toll of chronic stress and racism on the endocrine system, the environmental factors like heat and housing that disrupt hormonal balance, the intergenerational effects of trauma through epigenetic changes, and the biological costs of fatherhood, caregiving, and queer invisibility. At its core, *Under the Skin, Above the Pavement* asks: what happens to the body when masculinity becomes both an armor and a target?

This book is written for young men navigating survival, identity, and silence, as well as scholars, public health practitioners, and educators who seek to understand how systems of inequality are embodied. Blending narrative storytelling, cultural analysis, and scientific inquiry, it speaks to both those who live at the intersections and those who study these complex dynamics. There are many books about men, health, masculinity, and race. But this one brings them together with urgency and depth, providing a holistic view of how identity, environment, and biology intersect to shape the risks men face every day.

COMPARATIVE TITLES

1. *Heavy: An American Memoir* by Kiese Laymon (Scribner, 2018)

Laymon's powerful exploration of weight, masculinity, trauma, and survival in a Black Southern context shows how personal narratives can embody structural violence. *Under the Skin, Above the Pavement* shares this vulnerability and critique, but does so through a public health lens— drawing connections between embodied masculinity, biology, and urban systems.

2. *Dying of Whiteness* by Jonathan M. Metzl (Basic Books, 2019)

Metzl's exploration of how racialized politics harm public health resonates with the systemic critique in *Under the Skin.* However, *Under the Skin, Above the Pavement* centers men of color, masculinity, and the embodiment of risk, offering a perspective largely missing from existing public health narratives.

3. *Invisible Men: Mass Incarceration and the Myth of Black Progress* by Becky Pettit (Russel Sage Foundation, 2012)

Pettit's work uncovers the hidden consequences of incarceration on Black men's life chances. While Pettit's focus is sociological and statistical, *Under the Skin, Above the Pavement* expands the narrative by connecting incarceration and surveillance to biological tolls, such as hormonal dysregulation and epigenetic shifts.

4. *Black Men in a White Coat: A Doctor's Reflection on Race and Medicine* by Damon Tweedy (Picador, 2015)

Tweedy's memoir blends personal narrative and clinical insight to reveal how race and health intersect in medicine. While *Under the Skin, Above the Pavement* similarly blends science and story, it moves beyond the clinical realm to include epigenetics, environmental health, masculinity, and urban ecology.

5. ***Body and Soul: The Black Panther Party and the Fight Against Medical Discrimination* by Alondra Nelson (University of Minnesota Press, 2011)**

Nelson's historical account connects grassroots organizing to medical justice. *Under the Skin, Above the Pavement* builds on this tradition but shifts the lens from collective action to the individual embodiment of structural neglect; highlighting how marginalized men's bodies archive generational trauma and environmental exposure.

HOW THIS BOOK STANDS OUT

Unlike existing titles that focus primarily on sociology, history, or memoir, *Under the Skin, Above the Pavement* brings together: Public health research, Narrative storytelling, Environmental Science, and Critical masculinity studies. Its intersectional lens, rooted in lived experience and scholarly depth, makes it a first of its kind interdisciplinary text for both young men navigating identity and scholars advancing equity driven health discourse.

TARGET AUDIENCE
Primary Audience:

1. Young Men (18-35)

   - Focus: Urban Black, Brown, and marginalized men facing systemic inequality
   - These readers will see their lived experiences reflected in the book, whether it's navigating chronic stress, hyper-surveillance, environmental challenges, or identity struggles tied to masculinity. This group seeks a sense of belonging, validation, and a framework for understanding how their bodies bear the weight of systemic and societal pressures.

2. Scholars and Students

   - Focus: Public Health, Social Work, Sociology, Gender Studies, and Urban Studies fields
   - This group will be drawn to the interdisciplinary approach combining public health, environmental science, masculinity

studies, and critical race theory. *Under the Skin, Above the Pavement* offers a unique, academic yet accessible exploration of how race, class, gender, and environmental factors intersect to shape health outcomes.

3. Public Health Practitioners

- Focus: Professionals working on health equity, mental health, and environmental justice.
- These readers will appreciate the practical insights on urban health disparities, stress, trauma, and the social determinants of health. *Under the Skin, Above the Pavement* provides a framework for understanding how systemic oppression impacts both biological and social realities of health and offers ideas for addressing these disparities in practice.

4. Community Organizations and Activists

- Focus: Those working within marginalized communities, particularly around issues of health justice, environmental racism, and empowerment.
- This book serves as a resource for understanding how environmental and social systems affect community health, offering tools to advocate for and create healthier environments for marginalized populations.

Secondary Audience:

5. Gender Readers Interested in Health, Race, and Masculinity

- Focus: Readers interested in the intersection of race, masculinity, health, and identity in contemporary society
- These readers may be drawn to the book's storytelling approach, which provides not only scientific and sociological in-

sights but also personal and collective narratives, offering a humanizing perspective on complex issues.

6. Education and Policy Makers

- Focus: Teachers, activists, and policy influences, involved in public health education, particularly in urban or underserved settings
- This book offers insight into how education and policies can be shaped to address the intersectional impacts of inequality on men's health and well-being.

CHAPTER SUMMARIES

Prologue: The Weight We Carry, The Streets We Inherit

In the prologue, the author reflects on his upbringing in Charleston, South Carolina, tracing how race, masculinity, and environment silently shaped his health and identity. He introduces the central premise of the book: that the bodies of Black and brown men in urban America carry the cumulative weight of systemic neglect and cultural expectations. Blending personal narrative with a call to action, the prologue sets the stage for a book that seeks to bridge science and story, survival, and healing.

Chapter 1: The Biological Grind—Chronic Stress, Cortisol, Toxic Stress

This chapter sets the foundation by exploring how the urban environment contributes to chronic stress in marginalized men. It delves into the role of cortisol in the body, how toxic stress manifests biologically, and the long-term effects of living in environments that are marked by racial, economic, and social inequalities.

Chapter 2: Masculinity as a Risk Factor: Norms, Suppression, and Survival

Masculinity is not just a social construct but a biological risk factor. This chapter examines how societal norms around masculinity influence men's health, including the avoidance of help-seeking behaviors, higher rates of substance abuse, and delayed medical treatment, all of which exacerbate health risks.

Chapter 3: Masked in Plain Sight: Queer Masculinities and the Politics of Passing

This chapter introduces queer masculinities and the unique challenges faced by LGBTQ+ men, particularly in urban environments. It explores the emotional and physical toll of "passing" and the survival mechanisms necessary for those who navigate both homophobia and toxic masculinity. It reflects on the intersection of sexuality, identity, and urban stressors.

Chapter 4: Hood Epigenetics: Trauma, Memory, and Molecular Legacy

Building on the concepts of intergenerational trauma, this chapter explores how stressors such as violence, poverty, and discrimination affect gene expression. It discusses the emerging science of epigenetics and how historical trauma is biologically embedded, affecting the health outcomes of future generations of Black, brown, and marginalized men.

Chapter 5: Heat, Hormones, and Housing: Environmental Endocrinology in the Inner City

This chapter connects environmental stressors like heat, poor housing, and urban design with hormone disruption. It explores how urban inequality, inadequate housing, and climate changing intersect to alter the endocrine system, impacting health in profound ways, particularly for men in low-income neighborhoods.

Chapter 6: Urban Immunity: Food Deserts, Inflammation, and Resistance

This chapter highlights how urban environments contribute to chronic inflammation and suppressed immune responses. It examines the critical role of food access, dietary choices, and air quality in shaping immune system function, with a focus on how systemic inequities lead to heightened vulnerability to diseases like diabetes, malignant neoplasm, asthma, and hypertension.

Chapter 7: Concrete Wombs and Metal Cradles: Fatherhood, Care, and Inherited Wounds

This chapter explores the role of men in caregiving and fatherhood, both as biological and social imperatives. It discusses how caregiving is framed with masculinity, how trauma and resilience are passed down generationally, and how these roles intersect with men's health and wellbeing. It also considers how fatherhood impacts male identity and biological inheritance.

Chapter 8: From Streets to Studies: Reclaiming Research, Reclaiming Narratives

This chapter discusses the importance of community-driven research, emphasizing participatory methods that give voice to marginalized communities. It explores how data reclamation, especially in Black and brown communities, empowers people to tell their own stories, advance their own health agendas, and challenge institutionalized racism in public health research.

Epilogue: Street Medicine, Future Maps: Visions for a Healing-Centered World

The epilogue envisions a justice-oriented future where health equity is at the forefront of urban policy. It discusses how grassroots movements, street medicine, and community advocacy can transform public health and provide healing solutions rooted in the experiences of marginalized men. This future emphasizes justice, health, and healing as interconnected forces in the fight against systemic inequality.

Dedication and Acknowledgements

This book is born from countless conversations, quiet reflections, and the lived experiences of Black and Brown men navigating life within urban communities. It is dedicated to you — the brothers, sons, fathers, uncles, cousins, and friends who embody resilience, brilliance, and complexity in spaces that often fail to see the fullness of your humanity. You are the heartbeat of this work.

The initial spark for this book came from my dear friend, Ashley Glenn-Robinson. Our deep conversations about public health and social equity helped me begin to imagine how masculinity, health, and environment intersect in ways that shape the lives of so many men. Ashley, your insight, and spirit lives in every page.

To my circle of brothers Joshua Bryan, Joequise Wright, Colin Wagner, Stewart Dugger, Tyon Jones, Ricardo Robinson, Samuel Cooper, John Outlaw, Merrill Gadsen, Albert Shuler and Trevor Jones — thank you for the unfiltered talks about manhood, masculinity, education, biology, public policy, music, and public health. In our dialogue, I found a mirror reflecting how unique our stories are, yet how naturally they intertwine. Those exchanges helped me see that our differences are not divisions, but dimensions of understanding and that coexistence is both art and survival.

To my family: Kimberly Bass, Maurice Nelson, Shanice Nelson, Keon Nelson, Franchell Smalls-Lewis, Dexter Nelson, Regina Nelson, Lillie Nelson, and Bryce Nelson thank you for being the unseen hands that shaped my brilliance. Your love and patience held me steady through the long nights and quiet doubts. And to my extended family and friends who kept me accountable when I wavered, thank you for reminding me of the power of completion. This book is also a tribute to the Upward Bound and TRiO programs nationwide, institutions that reminded me that access and opportunity are not gifts, but rights. It taught me that success is possible when you believe in your potential, and I carry that lesson with me in every classroom and community I serve.

I would also like to express my gratitude to Dr. JaMuir M. Robinson, my dissertation committee member, for offering insight, patience, and grace during a pivotal chapter of my academic life. Your mentorship helped me translate passion into purpose. And to Coppin State University's College of Health Professions, the School of Allied Health, and the Department of Health Sciences, thank you for providing the foundation and environment that nurtured this work from an idea into scholarship.

Finally, to the readers, this book is not meant to challenge masculinity, but rather to invite reflection. It seeks to explore how masculinity is defined, shaped, and expressed within urban contexts, and how those definitions intersect with biology, psychology, and public health. My hope is that this work deepens our understanding of the men who too often bear the burden of societal misperception and reminds us that to see them fully is to heal more than just the individual it is to heal the community.

Contents

Prologue: The Weight We Carry	1
Chapter 1: The Biological Grind	4
Chapter 2: Masculinity as a Risk	21
Chapter 3: Masked in Plain Sight	39
Chapter 4: Hood Epigenetics	55
Chapter 5: Heat, Hormones, and Housing	69
Chapter 6: Urban Immunity	88
Chapter 7: Concrete Wombs and Metal Cradles	105
Chapter 8: From Streets to Studies	126
Epilogue: Street Medicine, Future Maps	149
About the Author	152
Notes	153

Prologue: The Weight We Carry

Prologue: The Weight We Carry, The Streets We Inherit

Growing up in Charleston, South Carolina, I watched the world move around me, never fully understanding how the weight of my skin, my neighborhood, and my family history shaped my existence in ways I couldn't see but could always feel. I lived in a world where survival was a daily practice, not just for me, but for every man around me, particularly those in Black, Brown, and marginalized communities. I saw men navigating a system that wore them down physically, mentally, and emotionally, but never in ways that were visible at first glance. It was in the quiet, subtle moments that the toll of trying to succeed within systems that were built for you to fail felt the heaviest. The gritted teeth during long shifts, the weary glances between brothers who understood that the weight they carried wasn't just their own. These moments, small, but resonant, are what I want to explore in *Under the Skin, Above the Pavement*. Not just the men I grew up with, but every man navigating a world designed to test his body, his mind, and his spirit.

The book you're holding isn't just about health disparities; it's about understanding the unseen forces that make those disparities feel inevitable. From the physical toll of stress that runs through our bodies like a constant undercurrent, to the cultural scripts we're expected to live by, I want to show how masculinity, race, and the very environments we inhabit are inextricably linked to how our bodies react, and how they break down over time. While the experiences of Black and Brown men in underserved urban environments share certain common struggles, they are far from uniform. Each man's journey is shaped by his unique identity, background, and intersectional experiences. This book aims to explore those differences while also acknowledging the shared realities that connect them. But this isn't just a story of survival. It's a story of inheritance, resilience, and reclamation. It's about the legacies we carry in our genes, our behaviors, and the way we're taught to hold everything in until it spills over. It's about how men are often expected to carry the weight of their worlds alone, without ever being taught how to share or release it. And ultimately, it's about breaking cycles of trauma, silence, and invisibility. As I reflect on my journey from that single-parent household to the classroom, I realize that the lessons I learned on the streets, in the halls of academia, and from the men around me can all be woven together into a story of hope that calls for transformation, healing, and the recognition of the full humanity of Black and brown men.

This prologue isn't just a prelude to a series of academic analyses; it's an invitation to see the world through the lens of those who have long been overlooked, misunderstood, or reduced to statistics. It's about understanding how systems of inequality, whether structural, cultural, or biological, collide in ways that impact health, survival, and identity. I'm inviting you to sit with the stories and the science, to understand how each chapter of this book can offer not only insight into the lived experiences of these men but also a path forward toward healing and justice. In many ways, this is the work I've always been meant to do. To bridge the worlds, I've walked through; between science and humanity, between struggle and hope, between surviving and thriving. My hope is that as you read these pages, you not only gain a deeper understanding of the intersections that shape health and identity but also begin to see that the way forward is possible. The body may carry the weight of generational burdens, but it can also heal, reclaim, and redefine itself. This is a story about survival, yes. But it's also a story about the future, and the future is ours to shape.

Chapter 1: The Biological Grind

THE BIOLOGICAL GRIND — CHRONIC STRESS, CORTISOL, TOXIC STRESS

"If you want to know how long someone will live, don't ask about their genetics, ask for their zip code." - Unknown.

Do you remember what happens when you go to the doctor? They check your blood pressure. Your height. Your weight. They measure your temperature, maybe draw blood for a metabolic panel, all to quantify your health using predetermined standards and critical values (although some scholars would convey some of these standards are outdated but that's another story for another day). While these measurements may offer a snapshot of your physical status, they miss something more profound. Because if we're being honest, the number that may best predict your health isn't in your lab results. It's in your address. Your zip code may be a more powerful predictor of your health than even your genetic code. A landmark study from Harvard University highlighted this startling truth, where you live can determine not only how long but how well you live.[1] This isn't just about proximity to hospital or the presence of sidewalks. It's about the entire social ecosystem: housing quality, school funding, air pollution, gun violence, green space, and economic opportunity. The Centers for Disease Control and Prevention (CDC) divides the entire social ecosystem into what public health experts would call the social determinants of health (SDOH).

Before we explore the biological grind in depth, it is important to define the framework through which these realities unfold. There are five domains of the social determinants of health as outlined by the U.S. Department of Health and Human Services that include:[2]

1. Economic Stability- employment, income, expenses, debt, and food security
2. Education Access and Quality- literacy, early childhood education, high school graduation, and higher education
3. Health Care Access and Quality- access to primary care, health literacy, insurance coverage, cultural and linguistic competencies
4. Neighborhood and Built Environment- housing, transportation, safety, parks, walkability, food access
5. Social and Community Context- social integration, support systems, community engagement, and exposure to discrimination

Not only does SDOH aim to characterize the area in which you work, play, eat, age, sleep, and die but their layered interactions can predict not only severity but susceptibility of health outcomes and health behaviors. According to some studies, environmental and social factors account for 80-90% of health outcomes, while direct medical care accounts for just 10-20%).[3] In short, health isn't just shaped in hospitals. It is shaped in neighborhoods. This chapter is about the biology of geography, the grind beneath the skin. It explores how the chronic stress of racial, economic, and social inequality in urban environments get under the skin of marginalized men and takes hold. For many Black and Brown men, the zip code becomes a diagnosis before a single test is run.

The Grind Beneath the Skin

The grind. It's a word that conjures both aspiration and exhaustion. For some men, it's a badge of honor. For others, it's a burden passed down through generations. For Black and Brown men in urban America, the grind is not a metaphor. It is physiological. It's biological. It's real. It's waking up before sunrise to hustle for a paycheck, caring for family, sidestepping violence (state-sanctioned or otherwise) and bracing from the microaggressions and injustices that meet them before breakfast. It's knowing that every room entered, every job applied for, every glance received potentially carrying the weight of history, bias, and survival. The grind begins long before disease has a name.

It starts in neighborhoods shaped by underinvestment, in schools expected to do more with less, and in workplaces where security is fragile and surveillance is constant. Sociological research on racialized stress shows that repeated exposure to discrimination and threat requires heightened vigilance in a state that becomes biologically ingrained over time.[4] That vigilance trains the body to expect danger even in moments meant for rest. The nervous system remains on alert, scanning for risk that may or may not arrive. Over time, this constant readiness interferes with recovery, sleep, and emotional regulation, a pattern well documented in stress and trauma research.[5] What appears as resilience from the outside often reflects a body that has learned it cannot afford to relax.

Because behind the grit is a physiology that is constantly on edge, adapted to stress but not immune to it. What begins as tension in the shoulders, or a racing heartbeat becomes, over time, a biochemical signature of living in a constant state of threat. Chronic activation of the body's stress-response systems alters how multiple biological systems function, a process described in neuroscience and endocrinology (the study of your body's chemical messengers (hormones) and the glands that make them) as prolonged stress dysregulation.[6] The grind gets beneath the skin. And once it's there, it doesn't just affect how someone feels; it affects how their body functions. It shapes everything from cardiovascular health to cognitive clarity. From sleep cycles to cellular aging. The grind is the pulse of inequality—steady, exhausting, and often invisible. Public health researchers describe the cumulative toll of this process as allostatic load, the wear and tear that results when the body is forced to repeatedly adapt to stress without sufficient recovery.[7] Elevated allostatic load has been linked to earlier onset of cardiovascular disease, metabolic disorders, immune dysfunction, and premature mortality.[8] These outcomes do not appear overnight. They accumulate slowly, often unnoticed, across years of sustained pressure. The body keeps track even when the mind pushes forward.

To understand how this grind operates beyond theory, consider the life of Nipsey Hussle. Born and raised in South Los Angeles, Nipsey was more than a rapper; he was a visionary artist and entrepreneur and community strategist who reinvested in the same neighborhood that shaped him. From the outside, his life appeared to embody transcendence—discipline, success, purpose, and control. He built businesses where disinvestment had been normalized and remained present where others were encouraged to escape. His story is often told as proof that individual determination can overcome structural constraint.But biology does not fully recognize success stories. It recognizes exposure. South Los Angeles, like many historically redlined neighborhoods, has long been shaped by environmental pollution, economic stress, neighborhood violence, and intensive policing patterns rooted in discriminatory housing and lending policies.[9] Even as Nipsey accumulated visibility and influence, his body remained embedded in an environment that demanded constant vigilance. Research on stress physiology shows that chronic alertness does not simply disappear with upward mobility; responsibility, visibility, and survival in high-stress environments carry their own physiological costs.[10] The grind shifts form, but it does not vanish.

Another way this wear becomes visible is at the cellular level through telomeres. Telomeres are protective caps (shaped similarly to the end of shoelaces) at the end of chromosomes that preserve DNA stability during cell division, a process central to aging biology.[11] Each time a cell divides, telomeres shorten slightly. When they become too short, the cell no longer functions normally and enters senescence or dies. Under chronic stress, this shortening accelerates. Studies show that prolonged exposure to psychosocial stress, including poverty, racial discrimination, and neighborhood disadvantage is associated with significantly shorter telomere, length, reflecting accelerated biological aging.[12] The process lies at the heart of what public health scholar Arline Geronimus describes as weathering, the accelerated biological aging that results from repeated exposure to social, economic, and racial adversity.[13]

Weathering helps explain why Black men often exhibit higher stress biomarkers and earlier disease onset than their White counterparts, even when health behaviors and socioeconomic status are similar.[10] Stress, in this sense, is not only experienced; it is biologically embedded. It accumulates quietly across years and, in some cases, across generations. The grind beneath the skin is not a personal failure. It is adaptation. Bodies adjust to environments that demand constant alertness by prioritizing survival over restoration. Heart rate stays elevated. Inflammation lingers. Sleep becomes shallow. Cells age faster than they should. What society often labels toughness; biology frequently experiences as threat. Understanding the grind at this level reframes illness not as weakness, but as information. Before we can fully examine, behavior, masculinity, or choice, we must first understand what chronic survival does to the body asked to carry it.

The Physiology of Stress: Cortisol and the Wear-and-Tear of Survival

Stress does not operate as a vague emotional state; it follows a biological chain of command. When the brain perceives threat, whether physical danger, financial instability, racial surveillance, or chronic uncertainty, it activates a coordinated survival response designed to protect the body. At the center of this response is the hypothalamic (in the brain)-pituitary (at the base of the brain) adrenal (located above the kidneys) [HPA] axis, the body's primary stress-hormone system. Once triggered, the HPA axis signals the release of cortisol, often dubbed as the "stress hormone" also mobilizes energy, sharpens attention, and prepares the body to respond. In short bursts, this system is adaptive. But when survival becomes the default rather than the exception, the same system that protects the body begins to strain it.[14]

Under chronic stress (whether psychological, environmental, or social), cortisol does not follow its normal rhythm of rising and falling. Instead, it remains elevated or becomes erratic, disrupting sleep, mood, blood pressure and emotional regulation.[6] This persistent activation contributes to the cumulative biological burden already described as allostatic load—the cumulative wear and tear/breakdown on the body's systems due to repeated stress leading to chronic diseases and other health problems.[15] For many men, this does not feel like crisis. It feels like being constantly "on edge," exhausted but unable to rest, emotionally muted yet internally tense. The stress-hormone system becomes the first point of strain, setting off a cascade that spreads across multiple biological systems.

One of the systems most affected by chronic stress is the cardiometabolic system. This system captures how things like blood sugar, insulin, cholesterol, and body fat directly affect your heart health and vice versa. Think of it as your body's power grid and delivery service; if the power (metabolism) is unstable and the delivery (circulation) is blocked, your whole system, especially the heart, suffers. Cortisol signals the body to conserve energy in anticipation of danger, increasing blood glucose and promoting fat storage. When stress is unrelenting, this survival adaptation alters how the body processes glucose and lipids. Chronic cortisol exposure increases insulin resistance and promotes abdominal fat accumulation; key pathways leading to cardiometabolic diseases such as diabetes, heart disease and stroke.[16] Research has shown that individuals with persistently elevated cortisol levels are approximately 25% more likely to develop cardiovascular disease and up to 50% more likely to develop type 2 diabetes compared to those with lower stress exposure.[17] This may help explain why men who remain physically active or "don't eat that bad" can still develop metabolic disease when stress is constant.

But the damage doesn't stop there. The immune system and inflammatory system are also reshaped by chronic stress. Inflammation is meant to be short-term and protective, activated briefly in response to injury or infection. Chronic stress, however, keeps the immune system in a low-grade inflammatory state. Over time, this persistent inflammation damages blood vessels, weakens immune regulation, and increases vulnerability to chronic illness.[18] Conditions such as asthma, autoimmune disorders, chronic pain, and increased susceptibility to infection often emerge from this biological terrain. For many men, this presents as feeling "run down," getting sick frequently, or living with ongoing pain without a clear injury. The brain and nervous system are not spared from the effects of prolonged stress exposure. Elevated cortisol affects regions responsible for memory, emotional regulation, and decision-making, including the hippocampus and prefrontal cortex.[19] When survival dominates the brain's priorities, long term planning, emotional processing, and self-reflection are deprioritized. A study published in *The Proceedings of the National Academy of Sciences* found that individuals with high cortisol levels were 30% more likely to experience cognitive decline as they aged.[20] In high-stress urban environments, these effects may begin much earlier, quietly limiting opportunity and increasing vulnerability to depression, impulsive, and substance use.

Taken together, these disruptions illustrate how stress becomes embodied across systems rather than isolated to any single organ. Cardiometabolic strain, immune dysregulation, neural reshaping, and hormonal imbalance interact, reinforcing one another over time. For men living in environments shaped by economic insecurity, racial discrimination, and structural inequality, this burden is magnified. [21] Studies show that Black men, in particular, are 1.5 to 2 times more likely than their White counterparts to exhibit elevated cortisol levels and other biomarkers or chronic stress.[22] Survival, in this context, is not a moment, it is a condition. This section anchors the biology of stress, but it does not stand alone. The next sections examine how toxic stress becomes racialized and spatial, how environments themselves function as chronic stressors, and how expectations of masculinity shape how men endure, suppress symptoms, and delay care. The body adapts to survive but adaptation has limits. Understanding those limits is essential before we can fully examine behavior, environment, and choice.

Toxic Stress and Structural Inequality

Toxic stress emerges when adversity is persistent, inescapable, and experienced in the absence of protective relationships or buffers. Unlike acute stress, which may be temporary and manageable, toxic stress rewires the body and brain over time. It changes how genes are expressed, how the heart beats, and how the immune system regulates inflammation. Urban environments saturated in inequality like red-lined neighborhoods, food deserts, food mirages, crumbling infrastructure, and discriminatory policing can create conditions where toxic stress thrives. These stressors are not accidental. They are the predictable result of historic and ongoing policy decisions that devalue certain communities. The body becomes the battleground where these structural forces are absorbed.

One of the clearest examples of this is redlining. In the 1930s (although Baltimore practiced this discriminatory practice in 1911), federal agencies like the Homeowners' Loan Corporation (HOLC) created color-coded maps to guide mortgage lending.[23] Green areas were deemed "best" and red areas (hence the name redlining) often where Black, immigrant, and low-income families lived were labeled "hazardous".[23] These red-lined neighborhoods were systematically denied access to capital, home loans, and insurance. The practice locked families out of wealth- building opportunities and led to long-term disinvestment. The consequences are not just economic, they are biological. Decades of underinvestment in redlined areas meant fewer parks, more industrial zoning, lower air quality, higher heat exposure, and increased police presence. These environmental conditions shaped the stressful landscape for entire communities. Redlining wasn't just a line on a map; it was a blueprint for embodied inequality. The stress created by these environments did not dissipate when policies changed on paper. It lingered in housing stock, neighborhood resources, and daily exposure to risk.

My own dissertation, *An Association of the Transgenerational Implications of Redlining and Obesity on Pediatric Type II Diabetes*, examined how this legacy continues to shape health outcomes in children. My findings suggest that children living in formerly redlined neighborhoods captured by neighborhood detraction elements experienced higher odds of obesity and T2D, not simply due to individual behavior but because of inherited exposure to structural disadvantage.[24] This is not simply history; it is health, passed from one generation to the next through environments, stress exposure, and biological adaptation.I saw this long before I had language for it. Growing up in Charleston, I remember growing up in Charleston, I remember my mom working multiple jobs. There were days when dinner came from the corner store or fast-food restaurants, when we slept in the same room to keep the heat in and the lights flickering more than they should have. At the time, I did not know what I was witnessing. Now I understand that it was toxic stress, settling not only in her body, but in mine as well. The tension in our home, the silence of sacrifice, the unspoken fear of unpaid bills did not stay abstract. It settled in our bones. Years later, a childhood friend would develop hypertension at 17, a full two decades before most people even get screened. We used to joke that we were "built for the hustle," wearing endurance like armor. But none of us ever asked what the cost of that hustle really was. Toxic stress teaches bodies to survive early, often at the expense of long-term health. What looks like resilience is frequently the body adapting to instability long before it has the chance to rest.

Toxic stress is not personal failing. It is the biological imprint of structural inequality. When exposure is chronic and recovery is scarce, survival becomes a lifelong posture. Understanding this process requires looking beyond individual choices and toward the conditions that repeatedly demand adaptation. To fully grasp how stress becomes disease, we must now examine how environments themselves act as stressors; how heat, pollution, and noise, and neglect transform survival into an everyday physiological burden.

When The Environment Becomes the Enemy

City life is not inherently unhealthy. But in communities where structural violence is layered with environmental hazards, it becomes hostile to health. Poor air quality, extreme heat, exposures, noise pollution, and overcrowded housing are not isolated inconveniences. They're stress amplifiers, layered onto lives already burdened by economic insecurity and racial surveillance. When these conditions persist, the environment stops being neutral. It becomes adversarial. Air quality offers one of the clearest examples of how places translate into biology. Research consistently shows that residents of historically redlined neighborhoods are more likely to be exposed to harmful air pollutants. A 2023 study by the American Lung Association revealed that residents of historically redlined areas are 2.4 times more likely to be exposed to harmful air pollutants, which are linked to asthma, cardiovascular disease, and early mortality.[25] These exposures are not evenly distributed. Communities that were systematically disinvested decades ago now bear disproportionate respiratory and cardiac burdens. Breathing becomes labor when the air itself irritates the lungs and inflames the cardiovascular system. Over time, chronic respiratory stress compounds the physiological effects of cortisol and inflammation already at work. These same neighborhoods can experience life expectancy gaps of 10 to 15 years compared to wealthier, predominantly white communities just a few miles away.[25]

Heat operates in a similar way. Formerly redlined neighborhoods tend to have fewer trees, less green space, and more heat-absorbing surfaces such as asphalt and concrete. As a result, these areas experience significantly higher temperatures during summer months, a phenomenon known as the urban heat island effect. A 2020 study in *Nature Communications* found that formerly redlined areas are, on average, 5 to 12.6 degrees Fahrenheit hotter during summer months due to the urban heat island effect, largely caused by reduced tree canopy and more heat-absorbing surface.[26] Such sustained exposure increases the risk of heat-related illnesses like stroke, dehydration, and respiratory distress, particularly among those with pre-existing conditions. Heat also disrupts sleep, further impairing the body's ability to recover from stress. When cooling becomes a privilege rather than a guarantee, the body pays the difference. Food environments add another layer of physiological strain. Many urban neighborhoods are characterized not only by food deserts, but by food mirages, areas where food is technically available but financially inaccessible or nutritionally inadequate. Diest shaped by limited options and economic constraint increase the risk of obesity, diabetes, and hypertension, but they also interact with stress biology. Chronic stress alters appetite regulation and promotes fat storage, particularly in the abdominal region. In this context, food choice is not simply behavioral; it is shaped by structural access and metabolic adaptation. The body responds to scarcity be preparing for more of it.

Noise pollution adds another invisible layer, but equally powerful, environmental stressor. Chronic exposure to traffic, sirens, construction, and neighborhood disorder keeps the nervous system in a state of heightened arousal. A 2018 study published in *Environmental Health Perspectives* found that residents exposed to chronic urban noise had cortisol levels 30% higher than those in quieter neighborhoods, even after controlling for socioeconomic factors. [27] These elevated stress hormones contribute to a range of negative health outcomes, from poor sleep, impaired concentration, and increased risk of metabolic and cardiovascular disease. Silence, like clean air and shade, becomes unevenly distributed.Housing conditions further intensify environmental stress. Aging infrastructure in disinvested neighborhoods increases exposure to mold, pests, and lead. According to the CDC, Black children are nearly three time more likely than White children to have elevated blood lead levels, largely dure to older housing stock linked to historic segregation and underinvestment.[28] Lead exposure affects neurological development, impulse control, and long-term health outcomes. According to the CDC, Black children are 2.8 times more likely to suffer from elevated blood lead levels compared to white children.[28] Lead exposure affects neurological development, impulse control, and long- term health outcomes. These early biological insults shape stress response systems long before adulthood, reinforcing cycles of vulnerability.

Environmental threat is not limited to physical exposures. Policing functions as a chronic psychological stressor in many communities of color. Surveillance, stop and frisk practices and the constant possibility of violent encounters with law enforcement condition residents to remain alert, cautious, and restrained. Studies have linked exposure to police violent (both direct and vicarious), to increased emergency room visits, anxiety, depression, and post-traumatic stress symptoms.

Black Americans are 2.5 times more likely to be killed by police—a statistic that extends beyond physical violence to mental health.[29] I remember walking past North and Payson in West Baltimore on a summer evening. It was humid, and the streets buzzed with movement, sirens in the distance, the scent of fried food in the air. A young boy couldn't have been older than nine glanced up from his bike when a police cruiser turned the corner. He didn't run or flinch. He just froze, eyes hollow, like he'd been taught not to move too fast. That stillness wasn't fear, it was conditioning. The boy wasn't growing up free. He was growing up prepared. This moment reminded me that the environment isn't neutral. It's an adversary some kids meet before they even learn to write their name.

Taken together, these environmental exposures: heat pollution, noise, housing conditions, food access, and policing do not act independently. They interact with stress biology, reinforcing allostatic load and accelerating wear across multiple systems. The environment becomes a constant signal to remain alert, to brace, to endure. For men navigating these landscapes, survival is not episodic. It is continuous. Understanding the environment as an active stressor reframes responsibility once again. Health disparities do not emerge solely from personal behavior or individual resilience. They are produced through sustained exposure to conditions that demand adaptation without offering recovery. When environments are structured to strain the body, illness becomes less a matter of chance and more predictable outcome. To fully understand the consequences of this exposure, we must now examine what resilience costs when survival is demanded day after day, especially from those taught never to show strain.

Unmasking the Daily Toll

The long-term effects of living in environments marked by racial, economic, and social inequalities are both sobering and systemic. Chronic exposures to these stressors does not just reduce quality of life, it shortens it. Studies show that allostatic load accumulates faster in people living under chronic adversity, leading to earlier onset of disease, reduced cognitive functioning, and premature mortality.[30] Life expectancy in some predominantly Black neighborhoods can be 20 years shorter than in adjacent white neighborhoods.[31] These disparities are not sudden or mysterious. They emerge from years of daily strain that often goes unnamed.Before illness becomes visible, stress embeds itself in everyday life. Sleep becomes fragmented. Fatigue becomes routine. Headaches, muscle tension, digestive issues, and irritability are normalized as the cost of getting through the day. For many men, particularly Black and Brown men, these symptoms are not interpreted as warning signs but as evidence of responsibility and endurance. The body adapts to constant demand, learning to function without recovery. Over time, this adaptation quietly erodes physiological reserve.

Beyond the physical toll of inequality, there's the psychological burden placed on men, especially Black and Brown men who are raised to internalize resilience as a form of manhood. This instruction to "stay strong" often translates into prioritizing work and family over personal health, which can lead to delayed doctor visits, underreported symptoms, and untreated conditions. The cultural script that defines manhood as stoic, self-reliant, and unyielding silences discomfort until the body can no longer compensate. These patterns show up clearly in how care is accessed. A 2024 study published in *American Journal of Men's Health* found that Black men were 43% more likely used the emergency department for usual care compared to their white counterparts, largely due to the societal pressure to appear strong and self-sufficient.[32] Emergency care becomes the point of entry not because men are unconcerned with health, but because seeking care earlier carries social, economic, and emotional costs.

For many men, missing work for a medical appointment is not a neutral choice. It represents lost income, disrupted family responsibility, and an admission of vulnerability that feels unsafe in environments where survival is already precarious. Health maintenance competes with economic necessity. Care is delayed until symptoms become unmanageable. By the time chronic conditions such as hypertension, diabetes, or heart disease are diagnosed, the biological groundwork has often been laid over decades. The normalization of pain as a form of resilience reinforces a dangerous loop: the longer you stay strong, the more likely you are to break. What is praised as toughness frequently reflects a body absorbing stress without relief. These delays in care are not simply individual decisions; they are shaped by structural barriers, masculine expectations, and environments that demand constant output with litter margin for repair.

The biological grind is real. It is the hidden link between cracked sidewalks and disrupted cortisol cycles, between over-policing and overactive immune responses. It is the connection between place and physiology, between racism and cellular stress. Unmasking the daily toll allows us to see how survival is carried in the body long before disease is named and why resilience, when demanded without support, quietly becomes a liability rather than a strength.

The Cost of Resilience
Resilience is often celebrated in communities of color, and it should be. Endurance, adaptability, and perseverance have sustained families and neighborhoods through generations of instability. But resilience can be also biologically expensive. The constant demand to "push through," to absorb trauma without visible cracks, comes with costs. The heroic narrative of the unbreakable man obscures the reality of hypertension, heart disease, and premature death. In many urban neighborhoods, manhood is forged in the fire of resilience, but it is a manhood built on stoicism, self-denial, and relentless endurance. There's no room to cry, no room to pause, no room to break or openly grieve, and even less permission to break.This cultural script of masculinity in these settings encourages boys to become men too

fast and expects those men to suppress pain indefinitely. Strength becomes synonymous with silence. Vulnerability is framed as weakness. Over time, the performance of manhood demands emotional restraint and physical overcompensation. But the body remembers what the performance of manhood tries to forget. Internalized pressure to remain strong and self-containment becomes cellular tension. The chest tightens. Sleep shortens. Blood pressure rises. These consequences are not theoretical— they are measurable, cumulative, and often fatal.

A qualitative study published in American Journal of Men's Health (2021) found that Black men who reported high adherence to traditional masculine norms were significantly more likely to report symptoms of anxiety, depression, and elevated blood pressure but far less likely to seek help.[33] In this way, masculinity does not merely shape behavior; it mediates how stress is embodied and how illness is delayed, minimized or even ignored. Population-level data reinforces this pattern. A 2020 report from the CDC showed that Black men have a 70% higher risk of developing heart disease by age 50 compared to their white counterparts, and a 1.5 times greater risk of experiencing a stroke in midlife.[34] These disparities are not just statistics; they are reflections of a system that requires emotional suppression, constant vigilance, and physical endurance as prerequisites for survival. Persisting in the face of chronic adversity is not free. The body keeps the ledger.

The cost of resilience deepens further when men are routinely exposed to death, violence, and distress. A research study noted that nearly 60% of Black men had witness a shooting before age 30.[35] Repeated exposure to trauma not only can have desensitizing effects, but it also deepens the stress footprint in the body, compounding the health toll.[36] Watching the disproportionate deaths of Black and Brown men, whether on the street or in the headlines doesn't just condition behavior; it conditions health. These deaths are not just statistics, they are somatic messages that shape survival, stress, and even generational biology. Resilience, in this context, becomes less a marker of strength and more a signal that suffering has been normalized. Reframing resilience does not mean rejecting it. It means recognizing its limits. Adaptation can keep people alive in the short term, but it cannot substitute for conditions that allow recovery. Strength should not require silence, and endurance should not demand self-erasure. Understanding the cost of resilience forces a harder question: not whether men are strong enough to survive, but why survival has been made so costly and so quiet in the first place? This chapter has traced how stress moves from structure to environment, from biology to disease and from survival to consequence. Because what happens *under the skin* begins *above the pavement.* The next step is to examine how masculinity organizes these responses, how beliefs about strength, responsibility, and worth shape behavior, delay care, and normalize pain. To understand health outcomes, we must now turn to the social scripts that teach men how to endure, and what they are taught never to say out loud.

Author's Note

As a scholar, as a son, and as a Black man who grew up in the echoes of redlining and resilience, this chapter is deeply personal. I wrote it to name the quiet violence, the kind that doesn't make headlines but still makes heart disease. This is not about pathology; it's about context. If we are to build a healthier world, especially for young man navigating both poverty and pride, we must interrogate the spaces that shape us before we ever step into a clinic. This is not just a story of bodies breaking down, it's a story of systems breaking trust.

Chapter 2: Masculinity as a Risk

Masculinity as a Risk Factor— Norms, Suppression, and Survival

> "The paradox is that masculinity, constructed as strength, becomes a mask that conceals fragility. And in that concealment, the damage multiplies."— bell hooks.

Masculinity is not just a social construct; it is a biological risk factor.[1] For many men, masculinity operates like armor: something worn to navigate a world that demands strength, control, and endurance. This armor teaches men how to carry pressure, how to keep moving when exhausted, and how to absorb stress without visible cracks. In environments shaped by instability, surveillance, or scarcity, this armor can be adaptive. It helps men survive. But what protects men socially can quietly endanger them over time.

For generations, boys have been taught that strength is measured by silence, that pain should be managed privately, and that asking for help signals weakness. These lessons are rarely delivered outright. They are communicated through praise, correction, ridicule, and omission of what is not said, often carrying as much weight as what is. Over time, emotional restraint becomes habit. Self-reliance becomes expectation. Vulnerability becomes risk. Research shows that many men internalize the believe that emotional control and independence are central to manhood, shaping how they cope with stress and whether they seek support.[2] The result is a kind of emotional armor, protective in the short term but corrosive over time.

Psychologists describe this tension through what is known as gender role strain. In plain terms, men are taught rules that are difficult, often impossible to meet all at once: be strong but never fail, be independent but always provide, be in control but never show fear.[3] This armor is often reinforced by social reward. Stoicism, toughness, and emotional containment are praised in families, workplaces, and peer groups. Men who endure without complain are seen as dependable, disciplined, and strong. But these same traits can make it harder to name distress, form supportive relationships, or recognize early signs of harm. Studies have found that men who strongly endorse traditional masculine norms are less likely to seek mental health care or preventive services, even when experiencing symptoms.[4] These patterns are not about lack of awareness. They reflect loyalty to a script that equates endurance with worth.

Importantly, the armor of masculinity does not emerge in a vacuum. For men navigating racialized or economically precarious environments, vulnerability can feel unsafe. Strength becomes a form of protection. Silence becomes strategy. Studies note that rigid masculine norms are associated with higher engagement in risk behaviors, lower help-seeking, and poorer health outcomes across the life course.[5] In this context, masculinity functions less as a personal choice and more as a learned survival response.

The armor itself is not the enemy. It exists for a reason. But armor is not meant to be worn indefinitely. Over time, it restricts movement, dulls sensation, and traps stress inside the body. The longer it remains in place, the harder it becomes to remove. Understanding masculinity as armor reframes the problem: men are not inherently resistant to care or emotionally unavailable. They are often following rules that once protected them and now quietly cost them. This chapter explores how rigid gender norms shape the health behaviors of men, often to their own detriment. From avoiding mental health support to delaying medical treatment, men are socialized to suppress vulnerability in the name of strength. By examining the interplay between cultural expectations and embodied health outcomes, we confront a hard truth: masculinity, as is often performed, is not a protective shield but a pathway to risk, not resilience.

The Social Scripts of Manhood

Masculinity is a performance rooted in repetition. From early boyhood men are handed scripts—unspoken rules about how to behave, what to feel, and what to suppress. These scripts dictate how to speak, how to stand, when to cry (never), and when to fight (whenever manhood is challenged). They come not only from family and peers but are deeply embedded in media, schools, religious institutions, and cultural narratives. Even when left unspoken, these codes are understood: to be a man is to be dominant, emotionally constrained, in control and capable of enduring pressure without complaint.

Sociologist R.W. Connell describes this process through the concept of hegemonic masculinity, which refers to the culturally dominant version of manhood that sets the standard against which all men are measured.[6] In simple terms, there is a "right" way to be a man, and most men cannot fully live up to it. Yet its influence is powerful precisely because it feels normal. Men may resist it, negotiate it, or fall short of it, but they are rarely free from its pressure. Hegemonic masculinity rewards dominance, emotional restraint, heterosexuality, and self-sufficiency, while marginalizing vulnerability, dependence, and difference.

What makes these scripts particularly effective is how they are enforced. Masculinity is policed not only by institutions, but by peers. Boys learn early which behaviors are praised and which invite ridicule. Emotional expression may be met with teasing. Sensitivity may be labeled weakness. Independence is rewarded; asking for help is questioned. Over time, these responses shape behavior. Men learn to anticipate judgment and adjust accordingly. Silence becomes safer than disclosure. Control becomes preferable to honesty.

When I asked some of my fraternity brothers what masculinity meant to them, the answers weren't uniform, but they weren't random either. Some said masculinity was about "being dependable," "handling your business," or "protecting your people." Others spoke of not folding under pressure," "having grit," or "leading without complaint." These responses echoed deeply felt values: strength, leadership, duty, and restraint. But almost no one mentioned vulnerability, emotional intimacy, or asking for help. Even in these well-intentioned definitions, you could hear the script: masculinity as control, performance, and endurance.

Importantly, these scripts are not evenly distributed or equally enforced. For men navigating racialized and economically constrained environments, masculinity is often intertwined with respectability and survival. Being seen as composed, controlled, and unbothered can feel necessary in spaces where scrutiny is constant and mistakes carry heavier consequences. In these contexts, masculinity is not simply about identity; it is about safety, dignity, and control. The cost of deviation feels high, and the margin for vulnerability feels narrow. Over time, men internalize these expectations so deeply that they begin to feel instinctual rather than learned. Masculinity stops feeling a script and starts feeling like truth. Men become both the performers and the enforcers, the prisoners, and the guards of the rules they were taught. For men who fall outside those norms, whether due to race, class, sexual identity, disability or emotional sensitivity, the penalties are steep. What began as social instruction becomes self-regulation. The armor hardens.

Understanding masculinity as a set of social scripts helps explain why behavior does not change simply through information or access. Men are not ignoring health messages; they are weighing them against rules they learned long before symptoms appeared. To change outcomes, we must first make those rules visible. Only then can we examine how coping strategies rooted in endurance and high effort begin to exact a cost, especially when survival itself becomes a lifelong performance.

Pushing Through: When Coping Becomes Costly
What happens to the body when you're never allowed to break down? When fear grief, and anxiety must be swallowed whole because masculinity demands it? For many men, especially those navigating racialized and economic marginalization, coping does not mean rest or recovery. It means endurance. Emotional suppression does not remain confined to the psyche; it becomes embodied, shaping how stress is carried day after day.[7] Chronic emotional repression, driven by masculine ideals of stoicism and control, has direct physiological consequences.[8] The nervous system cannot tell the difference between emotional pain and physical threat; both activate the body's stress response. When that activation becomes habitual, when men feel they must "man up" in the face of every emotional challenge, it produces wear and tear on the body.

This pattern of coping has been described as John Henryism, a high-effort strategy characterized by relentless determination in the face of chronic structural barriers.[9] In plain terms, John Henryism looks like pushing harder when things get harder. Stress is met with effort. Exhaustion is met with discipline. Vulnerability is met with silence. For many men, especially those raised in environments where slowing down feels unsafe or irresponsible, this approach is adaptive. It allows survival. But when high effort becomes the only acceptable response, coping itself becomes costly.

As introduced in Chapter 1, this cumulative toll of sustained stress is known as allostatic load, the biological price the body pays for constantly adapting to stress. In this chapter, the question sharpens: what happens when the stressor isn't just poverty or racism or environmental neglect, but masculinity itself? For men taught not to cry, not to talk, not to seek therapy or connection, stress becomes a permanent condition, an invisible hum beneath the surface of daily life. Black men, for example, have significantly higher allostatic load scores than white men by middle age, a difference linked not only to racism but to the cultural demand for emotional containment.[10] When masculinity valorizes suppression, it compromises the body's ability to recover.

One of the most visible consequences of this high effort coping pattern is hypertension. Black men develop high blood pressure earlier, experience it more severely, and 30% more likely to die from cardiovascular disease than white men, even when controlling for socioeconomic status.[11] These disparities cannot be explained by a lack of effort. In fact, the strategy of constantly pushing through, remaining vigilant, suppressing distress, and prioritizing responsibility over rest, which may accelerate stress-related disease. What is often praised as resilience externally may reflect continuous physiological strain internally.

Many men recognize this pattern intuitively. Chest tightness, chronic fatigue, poor sleep, irritability, and headaches are normalized as "just stress." Symptoms are minimized as long as work continues and responsibilities are met. The logic is simple: if you can still function, you keep going. But this logic has consequences. Elevated cortisol, persistent inflammation, disrupted sleep, and rising blood pressure are not random outcomes; they are embodied responses to prolonged suppression and effort. Over time, coping shifts from protection to harm.

John Henryism helps reframe men's health behaviors. Men are not failing to cope; they are coping intensely. They are using strategies learned in environments where vulnerability carried real risk. But strategies designed for short-term survival become dangerous when they are required indefinitely. Pushing through may get men past today's obstacle, but over time it converts stress into disease. Masculinity, when structured around endurance without release, transforms stress into illness quietly and predictably. The refusal to process emotional pain does not eliminate it; it embeds it. Chronic illness becomes the final language through which the body speaks. Understanding this pattern is essential before examining why men delay care, minimize symptoms, or wait until crisis to seek help—questions taken up in the next section.

Why Men Avoid Care (and Why It Makes Sense to Them)
If pushing through is the dominant coping strategy, then avoiding care is its logical extension. For many men, seeking medical or mental health support feels less like prevention and more like admission—admission that the armor has failed, that the body is no longer fully under control. Within dominant masculine norms, needing care is often interpreted as weakness, dependency, or loss of authority. Avoidance, then, is not accidental. It is consistent with what men have been taught about strength, responsibility, and self-reliance.

Health behavior research often frames care-seeking as a matter of awareness, access, or education. But for men, especially those strongly aligned with traditional masculine norms—the decision calculus looks different. Men are taught to weigh health risk through functionality: *Can I still work? Can I still provide? Can I still perform my role?* As long as the answer is yes, symptoms are minimized and care is deferred. This helps explain why men are significantly less likely than women to engage in preventive care, even when insured. National data show that men are 24-60% less likely than women to have visited a doctor in the past year, even after adjusting for health insurance coverage.[12]

Mental health care avoidance follows a similar pattern. According to the National Institute of Mental Health (2021), nearly 1 in 10 men have experienced depression, but fewer than half seek professional help.[13] In the U.S., men are less likely than women to receive mental health treatment, with only about 25% of men who report experiencing mental distress seeking any kind professional support.[14] This gap is not explained by lower need. Rather, it reflects stigma, masculine norms around emotional containment, and fear of being perceived as unstable or weak. For many men, distress is something to be managed privately, through work, distraction, substances, unresponsible sexual conquests, or silence rather than discussed openly or treated clinically. This disparity is even more pronounced among Black men, for whom cultural stigma and a history of medical mistrust create additional barriers to seeking care.

Avoidance does not mean disengagement from health altogether. More often, it produces delayed engagement. Men are more likely to enter the healthcare system through emergency departments rather primary care, often when conditions have escalated to crisis. Studies show that Black men, in particular are more likely to rely on emergency departments as a usual source of care, a pattern shaped by both structural barriers and masculine norms that discourage routine checkups.[15] Emergency care allows men to preserve the appearance of toughness until symptoms become unavoidable. Preventive care, by contrast, requires acknowledging vulnerability before collapse.

This logic becomes especially visible in men's avoidance of certain preventive screenings, most notably the digital rectal exam (DRE) used in prostate cancer screening. With the very nature of the procedure, requiring a doctor to insert a gloved finger into the rectum. While medically routine, the DRE collides directly with masculine norms surrounding bodily control, penetration, and vulnerability. For many men, the exam is not experienced simply as uncomfortable; it is symbolically threatening. It requires bodily exposure, passivity, and trust in another person—often another man—in ways that contradict dominant ideals of control and invulnerability. The procedure becomes loaded with fears of emasculation, loss of dignity, and social judgement.

Research suggests that men avoid prostate cancer screening not because they doubt its importance, but because the procedure conflicts with how masculinity has been socially constructed.[4] Among Black men, this avoidance is further shaped by historical trauma (buck-breaking) and medical mistrust, including legacies of bodily violation and exploitation. Fear of vulnerability is not abstract; it is informed by collective memory. In many communities, men reinforce one another's avoidance through joking, silence, or dismissal of screening as unnecessary unless symptoms are severe. The result is delayed detection and poorer outcomes, despite higher risk.

In the Chair: Calvin's Barber Shop

At Calvin's Barber Shop, the conversations flow easily. The space is familiar, masculine, and unsupervised by institutions. Men talk freely about work, money, relationships, and stress—often with humor acting as a shield. It is here, in the safety of the chair, that health anxieties surface indirectly. When prostate exams come up, the tone shifts. Jokes get louder. Laughter fills the gaps. "I ain't doing all that," someone says. Another adds, "I'll go when something's really wrong." The fear isn't framed as fear. It is framed as resolve. As boundaries. As control. No one says emasculation, but the meaning is clear. The exam is imagined not as prevention, but as violation. This rejection of prostate screening is a stark illustration of how masculine norms can literally undermine men's health and lead to preventable deaths. The impact of this stigma is particularly clear when we consider the higher mortality rates of Black men from prostate cancer, despite their being more likely to develop the disease at younger ages. According to the American Cancer Society (2021), Black men are at least twice as likely to die from prostate cancer as white men.[16]

What stands out is not ignorance. The men know prostate cancer exists. They know someone who has been affected. What they are negotiating is identity. To agree to the exam feels like crossing a line—conceding bodily control, risking judgement, and inviting vulnerability into a space where vulnerability is rarely permitted. The barbershop allows men to voice these concerns safely, but it also quietly reinforces the norms that keep them from care. Humor becomes both release and reinforcement. Silence follows.

The contrast between the barbershop and the clinic is instructive. In the chair, men feel seen and respected. In exam rooms, many anticipate dismissal or exposure. The barbershop becomes a confessional, the clinic, a threat. This difference helps explain why men may talk openly about stress while avoiding the very systems designed to address it. Economic pressure reinforces these patterns. For men in hourly, unstable, or physically demanding jobs, missing work for appointments carries immediate financial consequences. Health is deferred until responsibilities are met. Seeking care is postponed not because men do not value health, but because they are balancing survival trade-offs shaped by gender expectations and material constraint. Vulnerability is not only emotional, it is economic.

Trust also plays a critical role. Experiences of dismissal, discrimination, or disrespect within healthcare settings contribute to skepticism, particularly among Black men. Anticipation of being misunderstood intersects with masculine norms that already discourage disclosure. When men expect judgement rather than care, silence feels safer. Avoidance becomes protective. None of this suggests that men are indifferent to their health. Men avoid care because they are responding logically to a system of expectations that rewards endurance and penalizes vulnerability. Masculinity shapes how men interpret symptoms, assess risk, and decide when help is warranted. Avoidance, in this context, is not irrational, it is learned.

The cost of logic is cumulative, Conditions that might be manageable with early intervention (e.g., hypertension, diabetes, depression) progress quietly. By the time care is sought, treatment is more complex, outcomes are worse, and recovery is harder. Avoidance preserves masculinity in the short term but undermines health in the long term. Understanding why men avoid care requires moving beyond blame. Men are not failing to act; they are acting in alignment with the rules they were given. To change outcomes, those rules must change. Before that can happen, we must examine how those rules are unevenly enforced, how masculinity operates differently in spaces shaped by race, surveillance, and structural inequality. That is where we turn next.

Masculinity in the Margins: Urban Survival, Racialized Expectations

In urban environments, particularly in high-poverty, high-surveillance communities, and limited institutional support—masculinity is shaped not only by cultural norms but also by the politics of survival. For Black and Brown men, these environments often present daily challenges where their masculinity is tied to their ability to resist vulnerability and maintain and image of strength, toughness, and resilience. In neighborhoods, where poverty rates are high (nearly 20% of the U.S population lives below the poverty line, with Black and Latino communities disproportionately affected), masculinity becomes an essential tool for navigating systemic inequalities, including racial profiling, underemployment, and poor access to healthcare. The performance of masculinity here is defined by being the provider, the protector, and above all, the one who does not break, even in the face of overwhelming stress.

This adaptive performance has been described by Richard Majors and Janet Mancini Billson as the "cool pose," a culturally specific strategy through which Black men project control, confidence, and emotional distance in response to racial oppression.[17] Cool pose is not about bravado; it is about protection. Qualitative data from interviews and ethnographic studies reveal that many Black men in marginalized urban neighborhoods adhere to this hyper-masculine performance as a coping mechanism to deal with racism, economic disenfranchisement, and constant scrutiny.[17] However, it comes with serious health consequences. According to the American Psychological Association (APA), Black men are often taught that showing emotion is a sign of weakness, leading to higher rates of untreated mental health conditions compared to their white counterparts.[18]

These pressures are intensified by structural forces that shape daily life in urban communities. Mass incarceration plays a central role in defining masculinity for many Black and brown men. Black and brown men are incarcerated 5x the rate of white men; and more than 2.3 million people are currently incarcerated in the United States, a disproportionate share of whom are Black men.[19] Incarceration and surveillance become normalized features of male identity, sometimes framed as a badge of survival, other times as a permanent mark of exclusion. The effects extend beyond imprisonment itself. Formerly incarcerated men are 5x more likely to be unemployed than those without a criminal record, reinforcing economic instability and narrowing socially acceptable paths to masculinity.[20]

Moreover, the surveillance culture in these neighborhoods adds another layer to the performance of masculinity. In cities like Baltimore and Chicago, constant police presence, frequent stops, and heightened monitoring are part of everyday life for man young men of color. According to the American Civil Liberties Union (ACLU), Black men are 2.5 times more likely to be stopped by the police than white men.[21] This relentless scrutiny feeds into a sense of hypervigilance, where black men feel the need to embody a tough invulnerable persona at all times. This image of toughness becomes not just a response to social expectations but a necessary survival mechanism in environments where police violent is prevalent, and criminalization becomes a daily reality.

The psychological toll of this environment is substantial. Living under constant surveillance and threat produces chronic stress, anxiety, and emotional suppression. These conditions are associated with higher rates of untreated mental health disorders and stress-related illness among Black and brown men, including hypertension, cardiovascular disease, and depression.[18] Suicide reflects this burden starkly. Suicide has become a leading cause of death among Black men under the age of 35, with age-adjusted suicide rates among non-Hispanic Black Americans increasing by 53% between 2000 and 2020, including a rise among black men from 10.5 to 11.4 per 100,000 between 1997 and 2017.[22] Other estimates suggest a 25% increase in suicide rates among Black men over the past two decades.[23]

These trends begin early. Research indicates that Black adolescent boys experienced a nearly 60% increase in suicide attempts between 1991 and 2017, a pattern not observed in other racial or gender groups.[22] Early exposure to violence, policing, and constrained opportunity accelerates masculine socialization, teaching boys to suppress fear and manage distress independently long before they are given tools to process it. Growing up prepared becomes more important than growing up protected.

Understanding masculinity in the margins requires asking a different question. Rather than asking why men suppress emotion or delay care, we must ask what conditions make suppression feel necessary.

The behaviors examined throughout this chapter, pushing through, avoiding care, suppressing distress are patterned responses to racialized environments that demand constant self-regulation. Masculinity functions as a survival strategy that works socially, even as it exacts a heavy toll. In the next section, we turn to what happens when resilience itself becomes an expectation when strength is celebrated without acknowledging its cost. The question is no longer whether masculinity protects men, but what protection demands when it is the only option available.

Risk, Resilience and the Cost of Strength

Let's be clear: masculinity itself isn't inherently dangerous. The ability to persist, to protect, to lead are not pathologies. In fact, these traits are often adaptive and deeply necessary, especially in environments marked by instability, violence, or social neglect. Resilience is the story many men cling to with pride. But what happens when that same resilience becomes refusal, when the armor once worn to survive becomes too heavy to remove? The traits that support survival in crisis: stoicism, toughness, invulnerability often become liabilities in everyday life, where tenderness, vulnerability, and emotional expression are equally essential. Research supports this paradox.

Men who conform strongly to traditional masculine norms are significantly less likely to seek preventive care and mental health services.[24] According to the CDC, men in the U.S. are 50% less likely than women to engage with primary care services, and Black men are the least likely of all racial-gender groups to access mental health care.[25] These disparities are not simply behavioral; they are structural and cultural. Masculine socialization discourages men from acknowledging pain, admitting weakness, or prioritizing emotional wellness. In a national study, 61% of men agreed with the statement that "men should figure out their personal problems on their own without asking for help".[4] This mindset, while reflective of a deep sense of personal responsibility, also contributes to a pattern of untreated depression, stress related chronic illnesses, and even early mortality.

We can see the dynamic in the story of Jared, a 38-year-old Black man living in a predominantly working-class neighborhood of Charleston, South Carolina. Jared worked two jobs, supported his family financially, and was regarded as a dependable figure by his community. For years, he experienced chest tightness and chronic fatigue but dismissed the symptoms as stress and exhaustion— "part of the grind," he would say. When he finally sought medical care, he was diagnosed with Stage 3 hypertension and early signs of cardiomyopathy (chronic disease of the heart muscle). His doctor noted that the condition had likely been building for years. Jared admitted that he'd "always felt like going to the doctor was for people who had time to be sick." His experience is not uncommon. According to the American Heart Association (2024), Black men have among the highest rates of hypertension in the world, often developing it earlier and with more severe consequences than other racial groups. Jared's story embodies the silent cost of resilience, the way masculine pride, shaped by necessity, can turn into a slow, preventable unraveling.

The cost of strength, then, is not just measured in stress or suppressed emotion; it is measured in years of life lost, in family relationships strained, in silent suffering passed from one generation to the next. Black men in the U.S. have life expectancy nearly 4.7 years shorter than white men, and are disproportionately impacted by chronic conditions like diabetes, heart disease, and stroke many of which are preventable or manageable with early intervention.[26] But health doesn't just fail in isolation, it disrupts employment, increases caregiving burdens, and drives up medial debt. A 2022 analysis from the Milken Institute found that chronic diseases cost the U.S economy over $1.1 trillion annually in direct medical expenses and loss of productivity, with a disproportionate share of the burden falling on communities of color.[27]

For a low-income Black or Brown man, delaying care often means not just worse health, but missed days at work, job loss, and financial instability. Medical bills remain the leading cause of personal bankruptcy in the U.S., and men without stable insurance, especially those who are self-employed, gig workers, or underinsured often face impossible choices: seek care or keep the lights on. Jared's story, and the statistics behind it, make clear that masculinity's hidden costs are not just emotional or physical, they are deeply economic.

Public health has yet to fully reckon with masculinity as a social determinant of health. Masculinity is often viewed as background noise in conversations about race, class, and gender, rarely the subject of inquiry itself. But it should be. Masculine norms shape behaviors. Those behaviors shape bodies. And those bodies move through institutions such as clinics, hospitals, schools, and prisons, that reflect and reinforce the very norms that harm them. To dismantle the risk, we must name them clearly and reshape the systems that sustain them.

Reimaging Masculinity for Health and Healing
If masculinity can be a risk factor, then it can also be reimagined as a site of healing. We do not have to throw masculinity away to make men healthier, we must rebuild it, anchoring it in values that support not only survival but wholeness. A different script is possible, one where strength is expansive rather than restrictive, and where emotional honesty and community interdependence are a part of what it means to be a man.

A reimagined masculinity would recognize tenderness as strength, not its opposite. It would affirm that asking for help is not a betrayal of manhood but an act of courage. It would reject the idea that pain must be hidden or endured in silence. From a public health perspective, this shift has life-saving potential. Men who feel permitted to name distress, seek care early, and rely on others are less likely to reach crisis points that lead to emergency care, disability, or premature death.

There are already glimpses of this redefinition taking shape. Community-based interventions that promote emotional literacy, peer support, and culturally affirming care have shown promise in reaching men who are often disconnected from traditional health systems. Programs such as *The Confess Project* train barbers to become mental health advocates in Black communities, using trusted spaces to facilitate conversations around trauma, depression, stress, and self-care. Similarly, initiatives like *Fathers Incorporated* challenge narrow definitions of manhood by affirming caregiving, emotional presence, and vulnerability as central to masculine identity. These efforts do not ask men to abandon masculinity; they invite them to expand it.

Reimagining masculinity also requires attention to the environments that shape men's choices. Men cannot be asked to practice emotional openness in systems that punish vulnerability or deny time for care. Workplaces, schools, and community institutions play a role in reinforcing or disrupting the narrative that men must always be providers first and people second. When systems recognize men as whole humans with emotional lives, caregiving responsibilities, and health needs, healthier expressions of masculinity become more possible.

Ultimately, health equity for men, especially Black and Brown men require nothing less than a collective redefinition of what it means to be a man. It is not enough to reduce stigma; we must replicate it with a more generous, more truthful vision of masculinity, one rooted not in dominance or endurance, but in connection, care, and wholeness. Men's health does not depend solely on clinics, screenings, or interventions. It also depends on the stories men are allowed to live into, and the versions of themselves they are given permission to become.

Closing Reflection: What if Tenderness Were Survival?
What if we rewrote the story?

What if survival in our neighborhoods depended not on hardness, but on softness? What if the men who stood the longest didn't do so because they never broke, but because they learned how to bend? What if tenderness, so often policed, so often punished, was understood not as a liability, but as an evolutionary adaptation?

This chapter began with a difficulty truth: masculinity, as it's often performed, carries risk. But risk is not destiny. If social constructs can harm, they can also heal. Masculinity has been shaped by centuries of racial injustice, economic exclusion, and cultural silence. But it can be reshaped, by policy, by community, and by the intimate act of one man telling another, "You don't have to carry it all alone." Image a future where emotional expression was not a luxury but a norm. Where health clinics were designed not just to treat men, but to understand them. Where masculinity was not a performance to perfect, but a story to co-write, full of contradictions, full of tenderness, full of life. That kind of masculinity might lower blood pressure, reduce suicide rates, mend broken families. It might let men live longer, not just chronologically, but fully.

So, what if tenderness were survival?
Not a whisper of weakness, but the strongest thing we've ever been allowed to be.
Not the opposite of manhood, but the very thing that could finally set us free.
Because *under the skin*, where trauma lodges deep in muscle and memory, and *above the pavement*, where boys become men beneath surveillance and struggle, a different kind of masculinity is possible, one rooted in suppression, but in healing. One where being whole is the new form of strength.

Chapter 3: Masked in Plain Sight

Masked in Plain Sight—Queer Masculinities and the Politics of Passing

"All men live in a cage, but some men are forced to decorate theirs just to survive."- Darnell L. Moore, No Ashes in the Fire.

To be a queer man in America, especially one who is Black or Brown is to learn early that survival demands more than strength; it demands strategy. It demands silence. It demands the ability to shapeshift, to perform, to pass. Passing is not simply about sexuality. It's about safety. It's about the calculus made in front of the bathroom mirror each morning: *How deep should my voice be today? Should I say "bro" or "boo"?* Can I laugh freely, or will that sound like too much? This performance, rehearsed and revised over years, becomes muscle memory, so embedded that it's often mistaken for identity.

In many urban environments where being visibly queer can make you a target, many men learn to master invisibility. But masking one's truth does not come without a cost. The constant suppression of identity, affection, and authenticity doesn't just fracture the soul, it alters the body. It changes heart rates. It disturbs sleep. It escalates cortisol. The mask may protect from violence, but it exposes the wearer to slow, quiet harm.

This chapter lifts the veil on that performance. It explores how queer men, particularly those racialized and marginalized, navigate masculinity in spaces that ask them to be everything but themselves. It is a study of resilience, yes, but also reckoning with the psychological weight and biological toll of being masked in plain sight.

The Performance of Passing

To pass is to be read as something you are not, on purpose, for protection. In the context of queer masculinity, "passing" often means performing a version of straight, cisgender masculinity that is convincing enough to avoid ridicule, exclusion, or even violence. It is daily choreography, curating one's voice, walking, laughter, language, and gaze, not for style, but for survival. For many queer men of color, this performance is not simply about hiding their sexuality; it's about navigating layered threats in spaces that demand conformity and punish deviation. This performance is learned early. Schools, barbershops, locker rooms, church pews, and family cookouts become testing grounds.

Boys learn which movements are acceptable, which tones invite correction, and which expressions provoke danger. A wrist that bends too freely, a laugh that lingers too long, an interest that reads as "soft" can trigger ridicule or threat. Over time, these corrections accumulate into self-policing. The body becomes trained to anticipate judgment before it arrives. Passing, then, is not a single act but a sustained form of labor, a constant adjustment in response to imagined and real consequences. The burden of passing is well documented. A report from the Human Rights Campaign (HRC) found that 48% of Black LGBTQ+ youth felt they couldn't be themselves at home, and 55% reported being verbally harassed in school because of their identity.[1] These experiences do not disappear in adulthood. They simply migrate into new spaces such as workplaces, social settings, healthcare environments—where the stakes shift but the calculation remains.

Research shows that LGBTQ+ men who routinely conceal their sexual identity report significantly higher levels of psychological distress, depressive symptoms, and somatic complains than those who are able to live more openly.[2] Passing may reduce immediate exposure to violence, but it intensifies internal strain.For queer men of color, passing is complicated by racialized expectations of masculinity. It is not enough to avoid queerness; one must also embody control, composure, and emotional restraint in ways that align with dominant ideas of Black and Brown masculinity. A performance that succeeds in one context may fail in another. A man may pass at work but not in his neighborhood, in white spaces but not among other Black men, in public but not at home. There is no stable script, only constant recalibration.

Dewayne, a 33-year-old successful and educate Black queer man living in Windsor, North Carolina, does not pass easily. He has a large frame, a gentle voice, and expressive gestures that many could read as feminine. "It wasn't white folks that hurt me most," he explains. "It was Black men who looked like me, called me brother when it was convenient, then made me feel like something to mock, or worse, erase." In many environments, especially "Black spaces," Dewayne began to anticipate rejection. "I didn't walk into a room anymore; I braced for it. I clenched my jaw before I spoke. Not to seem 'man enough' but to prepare for whatever was coming: side eyes, slurs, silence. The most painful part wasn't the ridicule; it was the rejection—It is hard to explain," he says. "It's like being punished for something you didn't choose. Like your softness is a threat to their survival." That exclusion calcified into anger. Dewayne carried a deep resentment towards other Black men, believing for years that they were his greatest danger. "I hated them for not loving me. For needing me to disappear so they could feel okay." But healing began when

Dewayne found spaces that celebrated, rather than censored, his femininity. He found friends and fraternity brothers that didn't ask him to shrink. "They didn't flinch when I cried." Today, Dewayne says his femininity is no longer something he hides, it's something he honors.

Marcus, a 31- year- old from Washington, D.C tells a different but equally piercing story. He had long performed a version of masculinity that won him proximity to acceptance. "I'm everybody's boy, until I remind them, I'm not like them," he says. "At parties, I offered style advice, brought the energy, and held secrets. But the second my queerness became visible/the mask slipped, so did the welcome." Marcus recalled a moment that changed everything in a group of friends that he knew from elementary school throughout college. When a homophobic joke was made, Marcus waited, waited for someone to say something—anything. No one did. "That silence told me everything," he says. "I learned right then: I'm only safe when I'm useful. My dignity has an expiration date." Marcus still shows up for his community, but the heartbreak taught him to recalibrate. "I still ride for Black folks. I just had to learn who rides for me."

These stories reveal the emotional taxation of passing and the cost of failing to pass. Whether performing masculinity convincingly or being punished for not performing it at all, queer Black and Brown men are forced to navigate a narrow corridor or acceptability. The performance may offer momentary safety, but it also fractures belonging and isolates connection. Over time, the labor of self-monitoring becomes embodied. Even in moments of quiet, the body remains alert, waiting for the next correction.

Passing is often framed as a person choice. In reality, it is a response to structural conditions that reward conformity and punish deviation. It is an adaptive strategy developed in hostile environments. But strategies built for survival are not designed for rest. When passing becomes a permanent requirement rather than a temporary shield it constrains the self and erodes the possibility of ease. What remains is not simply a performance, but a wound rehearsed daily quiet, disciplined, and largely unseen. This performance sets the stage for deeper contradiction explored next. Because even when queer men pass, safety is never guaranteed. Masculinity does not simply demand conformity; it polices it. And for those who live at the intersection of race, gender, and sexuality, the margin for error is vanishingly small.

Queer but Not Safe— Masculinity as a Double Bind

Passing offers no guarantee of safety. It only postpones punishment. This is the double bind at the center of queer Black and Brown men's lives: masculinity is required for survival, yet full access to masculinity is never granted. Perform it too visibly, and you risk being labeled aggressive or threatening. Perform it too softly, and you risk ridicule, exclusion, or violence. There is no correct performance, only constant evaluation. For queer men of color, masculinity operates as conditional membership system. One is expected to carry the weight of manhood (strength, responsibility, restraint), without being fully recognized as a legitimate man. To be queer is to be seen as failing masculinity, even when one is actively trying to uphold it. This contradiction forces men into an impossible position: they must perform masculinity to remain safe, while knowing that masculinity itself is the standard used to mark them as suspect.

This bind is often enforced early, long before sexual identity is named. I remember being seven years old in Charleston, South Carolina and walking a few steps ahead of my dad toward the entrance of a Walmart. He called out, not unkindly but firmly, "Boy, put your hands down. Don't sway your arms as you walk." I froze. In that moment, I realized my body was being watched, not just by my father, but by the world he knew would be less forgiving. It wasn't just a correction; it was a warning. A defense. A lesson passed down, not out of hate, but out of fear. My father was trying to protect me from a society that would punish me for moving through it too freely. As I grew older, I became fluent in the choreography. I lowered my voice. I learned not to clap "too excitedly," not to laugh too loudly in mixed company, and not to linger too long when complimenting another man. Every gesture became a calculation. Every outfit, a decision. I wasn't trying to pass as straight; I was trying not to be targeted. And beneath the self-policing, resentment brewed, not just at the world but sometimes, heartbreakingly, at my dad and other Black men who enforced the very norms that excluded me.

As boys grow into men, the rules expand but the logic remains the same. Masculinity is policed through ridicule, silence, and the threat. Sociologist C. J. Pascoe describes how "fag discourse" functions less as a statement about sexual orientation and more as a mechanism for enforcing masculinity itself.[3] Boys are labeled, disciplined, and corrected for any deviation from expected masculine behavior—emotional expression, aesthetic difference, gentleness, or vulnerability. The language may shift with age, but the policing does not end. It moves from schoolyards into offices, barbershops, churches, and family kitchens, where men continue to audit their voices, gestures, and silences.

The double bind is also historically constructed. Throughout American history, masculinity has been defined in opposition to femininity and queerness. The hypermasculine image of the Black man, stoic, strong, aggressive was not just socially encouraged; it was weaponized by media, politics and even within some community norms as a defense mechanism against white supremacy. Queer Black men are positioned as liabilities, accused of weakening the image meant to protect the group. Yet queer Black men have always existed, contributing profoundly to culture, politics, and resistance. Figures like James Baldwin, Marlon Riggs, Bayard Rustin were celebrated for their brilliance while being quietly disqualified from full recognition as men because they disrupted the narrow script.

Even within LGBTQ+ spaces, the double bind persists. Masculinity is often fetishized, while femininity, especially in men of color is marginalized. "Masc for masc" is not just a dating preference; it is a social hierarchy that reproduces the same logic of exclusion found outside queer spaces. Queer men who do not, cannot or will not perform a heteronormative version of masculinity often find themselves excluded from spaces that promise liberation but replicate familiar punishments. This rejection, compounded by racism and classism, leaves many queer men of color stranded between communities, belonging to both and neither. Belonging remains conditional.

Psychologically, this bind manifests as hypervigilance and chronic self-surveillance. Men learn to monitor themselves not just for safety, but for acceptability. The question is no longer *Am I safe?* but *Am I allowed?* Over time, the effort required to manage this contradiction becomes exhausting. Survival tactics harden into identity. It becomes difficult to distinguish which parts of the self are authentic and which were edited to remain intact. What makes the double bind especially corrosive is that it isolates. Queer men may come to see straight men as their primary threat, while straight men are taught to distance themselves from queerness to protect their own standing.

Masculinity thrives on this division. It convinces men that safety lies in policing one another, rather than questioning the system that demands constant performance in the first place. The double bind is not simply a tension; it is a trap. A structure that forces men to choose between authenticity and safety, belonging and truth. And the longer one lives inside it, the more invisible it becomes. The next section turns to the body itself and how these judgements, exclusions, and contradictions are written onto flesh, shaping who is seen as desirable, acceptable, or disposable.

Bodies That Don't Belong

The body is never just a body. It is a social text, read, interpreted, and judged long before a word is spoken. For queer Black and Brown men, the body often becomes the primary site where masculinity's contradictions are enforced. It is where belonging is decided and denied. How a body moves, how it carries weight, how it gestures, how it takes up space—these cues determine whether a man is perceived as acceptable, suspect, or disposable. In spaces governed by rigid masculine norms, certain bodies are granted legitimacy while others are marked as excessive or deficient. A body that is too soft, too expressive, too fluid, or too large is often read as failing masculinity. These readings are not abstract. They occur in locker rooms, on sidewalks, in classrooms, on dating apps, and in family homes. Surveillance is constant, and it is rarely subtle.

Research reflects what many men already know. Sexual minority men report significantly higher levels of body dissatisfaction than their heterosexual counterparts, with the burden intensified among Black and Latino gay and bisexual men.[4] In one national study, nearly 80% of gay men reported being dissatisfied with their bodies- compared to only 45% of heterosexual men.[5] These disparities are not simply about appearance; they are about safety, desirability, and access to belonging.

Queer bodies are often required to perform contradiction. They must be visible enough to be legible yet restrained enough not to provoke threat. In heterosexual spaces, queer men's bodies are frequently framed as dangerous, deviant, or weak. In queer spaces, those same bodies may be ignored, fetishized, or sorted according to narrow hierarchies of masculinity. Softness becomes a liability. Fatness becomes disqualifying. Femininity becomes grounds for erasure.

Fatness, in particular, operates as a boundary marker. Dating apps like Grindr, Jack'd, and Scruff allow users to filter by height, weight, and body type, an algorithmic shorthand for exclusion. Studies indicate that nearly 40% of LGBTQ+ men who identify as overweight or fat report experiencing stigma in queer spaces, including rejection, ridicule, or invisibility.[6] Additionally, the "Bearspace" study conducted by the University of Brighton found that many fat gay, bi, and queer men felt uncomfortable and self-conscious in mainstream LGBTQ+ spaces due to anti-fat stigma, often feeling rejected or marginalized because of their size.[7] Furthermore, QX Magazine also reported that plus-size men in the LGBTQ+ community face significant stigma, with many feeling uncomfortable or out of place in spaces that prioritize thinner bodies.[8] Club scenes and ballrooms that once promised liberation can also replicate hierarchies that prize chiseled bodies and exclude softness, literal and figurative. The message is clear: to be large is to take up "too much" space, to embody too much presence, too much vulnerability, too much deviation from control.

Consider the story of Jamarr, a 30-year-old Black gay man from Atlanta. He describes the pain of being visible and invisible at the same time. "In straight spaces, I'm seen as a threat. In gay spaces, I'm either ignored or fetishized. Either I'm too soft, too fat, or I'm a 'type,' not a person. No one just sees me." His body, a map of resilience and rejection does not fit into the curated aesthetics of queer masculinity. Research supports this experience: weight-based discrimination was significantly correlated with internalized homophobia and self-esteem challenges among queer men of color.[6]

Family spaces often reinforce these lessons early. Many queer men learn early that their bodies are a disappointment long before they are understood as sexual beings. A laugh that is too loud, a walk that is too fluid, legs crossed the "wrong" way—each correction teaches the same lesson: your body is a problem. Over time, this policing trains men to mistrust their own physical presence, leading to chronic self-monitoring and bodily shame.

What makes this especially painful is the contradiction between rejection and appropriation. In fashion, dance, music and pop culture, the aesthetic innovations of queer men, especially Black and Brown queer men are regularly copied and commodified. Voguing, ballroom culture, and even slang like "clock that," "read," "gag" and "shade" have been pulled into the mainstream without honoring the people who created them. This contradiction breeds a deep dissonance: being punished for who you are, while watching the world profit from your reflection.

To live in a body that does not belong is to live in a state of constant negotiation. Every space requires calculation. Every movement carries consequences. The question is never simply *Who am I?* But *how will my body be read today?* This relentless evaluation does not end when the lights go out; it follows men home, into sleep, into intimacy, into health. The next section turns to place—to the neighborhoods, households, churches, and barbershops where these bodies must be managed most carefully. It is there, in the spaces called home, the hiddenness becomes not just strategy, but necessity.

Home, Hood, and Hiddenness

For queer Black and Brown men, "home" is often not a sanctuary. It is often a site of negotiation, silence, and strategic restraint. The very spaces meant to offer safety—family living rooms, neighborhood sidewalks, church pews, and barbershops can become environments where identity must be carefully edited. Hiddenness is not a preference; it's a survival strategy shaped by geography, surveillance, and the consequences of being read incorrectly. The neighborhood or "the hood," imposes a masculinity shaped by scarcity, surveillance, and structural oppression. Respect is currency, toughness is armor, and deviation is dangerous. Research shows that Black LGBTQ+ men living in high-poverty, high-policing neighborhoods are more likely to conceal their sexual identity due to fears of violence, job discrimination, and social ostracization.[9] Concealment may reduce immediate risk, but it amplifies psychological distress, depression and substance use over time.[10]

The hood teaches early lessons. Walk with confidence, but not softness. Speak firmly, but not freely. Be seen, but not too visible. Masculinity here is not abstract, it is enforced through observation and rumor. A look held too long. A gesture misread. A tone questioned. Hypervigilance becomes muscle memory. Few spaces capture this tension more clearly than the barbershop. Often celebrated as a cornerstone of Black masculinity, the barbershop is where boys learn how to become men, so long as they perform the "right" kind of masculinity. For many queer men, it is also where self-editing begins. Lower the voice. Laugh at the joke that stings. Avoid mentioning who you love. It's not just a haircut; it's an audition. This is a reminder that visibility comes with risk. Masculinity is affirmed, but only conditionally.

The same is true of churches. While the Black church has been a source of resilience and a cornerstone of cultural identity and political mobilization, it has also been a site of painful contradiction and erasure. The pew may welcome queer men for their musical gifts or organizing skills, but not for their full selves. A 2021 Pew Research survey reported that only 31% of Black LGBTQ+ adults feel accepted by their religious community.[11] In sermons and whispers, they are told to "pray the gay away" or to "walk in deliverance." The message is clear: be who we say you should be or be invisible. Thus, love is offered with conditions. Silence is framed as salvation.

Hiddenness often follows men into intimacy and identity itself. This is where the "down-low" phenomenon emerges, not as deception, but as survival. Men who have sex with men (MSM) without openly identifying as gay or bisexual are navigating layered risks: community violence, family rejection, economic precarity, and institutional stigma. Public health data show that men who have sex with men, including those who identify as straight, or DL are disproportionately affected by HIV. In fact, Black MSM account for 26% of all new HIV diagnoses in the United States, despite making up a small percentage of the population.[12] Among those, 1 in 2 are projected to be diagnosed with HIV in their lifetime if current trends continue.[12] These patterns are not about individual behavior or recklessness; they are about systemic stigma, lack of culturally competent healthcare, and the social cost of visibility.

And then there is the family home. The living room where uncles crack jokes about "soft boys." The kitchen table where a cousin or church member gets called out for "acting funny" or being a "fairy." For many, homes become a tightrope: too queer and you lose affection; too hidden and you lose yourself. Some manage this by splitting identities, out in the club, closeted in the crib. Others leave entirely, creating chosen families in cities miles away from the ones that raised them. Yet even distance does not erase longing. The grief of partial belonging lingers.

Hiddenness is not just social, its emotional, spiritual, and biological. Chronic identity suppression is linked not only to cardiovascular strain and immune dysregulation, but also to sleep disorders, disordered eating, chronic inflammation, and anxiety-related gastrointestinal conditions.[13] The body absorbs what the mouth cannot say. Every swallowed truth tightens the chest. Every edited gesture drains energy. Over time, concealment becomes exhaustion.

And yet, hiddenness is not the end of the story. Within these constrained spaces, men find ways to breathe. After-hours conversations. Side glances that say *I see you*. Quiet acts of defiance. Even survival strategies contain seeds of resistance. The next section turns to those margins, not as places of exile, but as sites of possibility where queer Black and Brown men begin to rewrite masculinity on their own terms.

Liberation in the Margins

Liberation, for queer Black and Brown men, rarely arrives through institutions that once demand their silence. It emerges instead in the margins, in spaces carved out of necessity, creativity, and refusal. These are not utopias, and they are not free from harm, but they are places where masculinity loosens its grip and new ways of being take shape. In these margins, men are not asked to perform invulnerability. They are allowed to breathe, to soften, to be contradictory. Liberation here is not the absence of struggle; it is the presence of choice.

For some, liberation requires distance. Jalen, 27, grew up in Florence, South Carolina, where masculinity was enforced through discipline, silence, and threat. He knew he was gay by middle school, but he also knew saying it aloud would cost him safety. "My dad believed in being a man," Jalen explained. "Cut grass. Lift weights. Be interested in cars. Don't let nobody punk you. Every time I laughed too loud or liked something different, I felt him watching." Home became a place of constant calibration, too queer meant rejection, too quiet meant erasure. Eventually, Jalen left. Moving to Houston didn't erase the grief of family distance, but it gave him room to exist without shrinking. "I found people who didn't ask me to edit myself," he said. "But I still had to grieve the version of home that couldn't love all of me." Liberation, for Jalen, came through leaving and through morning what staying made impossible.

For others, liberation does not come from departure, but from reframing. Jordan, 45, a Black gay attorney from Detroit, described a slower, internal shift. "For years, I thought I had to choose, be Black or Be gay," he said. "The hood made it clear there wasn't space for both." Over time, Jordan began to unlearn that lie. He found community not by disappearing, but by refusing to fragment himself. "Once I stopped asking permission to exist, everything changed," he explained. "My queerness stopped feeling like a wound. It became a compass." Jordan's liberation did not require escape; it required redefining masculinity on his own terms, integrating softness, authority, and care without apology.

These stories reveal an essential truth: there is no single pathway to freedom. Liberation looks different depending on age, class, geography, and risk. What unites these experiences is not where men end up, but what they refuse to carry forward. In the margins—ballroom spaces, chosen families, affirming barbershops after hours, community healing circles—queer men are not anomalies. They are architects. These spaces function as laboratories of possibility, where masculinity is not erased but expanded, no longer rooted in domination or denial, but in connection and truth. Importantly, liberation in the margin is not only personal— it is biological. Research shows that LGBTQ+ people of color who participate in culturally affirming spaces report lower levels of internalized stigma, reduced depressive symptoms, and stronger social connectedness.[14] These environments do more than health emotionally; they interrupt the physiological toll of chronic vigilance. They offer rest to bodies long conditioned for defense.

Liberation, then, is not about visibility for its own sake. It is about safety, sustainability, and wholeness. It is about creating spaces where men are not punished for softness or required to perform strength at the cost of their health. In the margins, queer Black and Brown men are not waiting to be accepted into masculinity—they are rewriting it. And in doing so, they offer a blueprint not just for queer survival, but for collective healing.

Conclusion: What if Visibility Were Safety?
What if the most dangerous thing a man could do wasn't to fight, but to be fully seen?

For queer men, especially those who are Black and Brown, visibility has long been a paradox. To be visible is to be legible, affirmed, and whole. But it has also meant exposure to ridicule, rejection, violence, and erasure. In many urban spaces, visibility has been a risk, not a right. Safety has required silence. Masculinity has demanded concealment. Survival has depended on knowing when to disappear. Yet this chapter reveals a deeper truth: queer men are not uniquely burdened by masculinity. They are simply forced to confront it sooner, more explicitly, and often at greater cost. The performance of passing, the policing of bodies, the pressure to suppress softness—these are not queer problems alone. They are symptoms of a system that teaches all men that vulnerability is dangerous and that worth is measured in control.

Heterosexual Black and Brown men are taught many of the same lessons. Do not cry. Do not flinch. Do not ask for help. Do not show fear. The scripts may differ in tone. But the message is very much the same: visibility invites punishment. Silence ensures survival. And so, men (queer and straight alike) learn to armor themselves. They perform strength. They swallow pain. They endure. And their bodies pay the price.

What queer men expose, simply by existing, is the cost of that armor. Their lives make visible what masculinity tries to hide: that constant performance fractures the self; that suppression accumulates biologically; that living in defense mode erodes health, intimacy, and longevity. Queer men are often framed as exceptions to masculinity, but in truth, they are its canaries; revealing what happens when authenticity is treated as threat. This is why visibility matters. Not as a spectacle. Not as defiance for defiance's sake. But as interruption.

What if visibility were safety, not because danger disappeared, but because masculinity loosened its grip?

What if a boy could walk through this neighborhood with softness and certainty and be met with care, not correction?

What if a man's tenderness was read not as weakness, but as intelligence?

What if being seen did not require being strong first?

Because at the core, queer and heterosexual men are not opposites. They are shaped by the same system, one that equates manhood with suppression. Both are told to carry pain quietly. Both are taught that intimacy is risky. Both are rewarded for endurance and punished for honesty. And both suffer the consequences in their bodies, their relationships, and their shortened lives. But this chapter also shows that another masculinity is possible.

In the margins, men are already building it. In chosen families. In affirming spaces. In moments where someone says, "You don't have to carry that alone." Queer men, having survived the demand to disappear, offer a blueprint for something different—how to bend without breaking, how to feel without collapsing, how to build belonging from fragments. Straight men, if they choose, bring access, protection, and power not to police difference, but to dismantle the conditions that make hiding necessary. Visibility, then, becomes more than exposure. It becomes strategy. A way to reimagine streets, barbershops, homes, churches, and clinics as places where difference is not punished but practiced. Where Black and Brown men of every identity are allowed to show up not as performances of strength, but as full, complicated, living selves.

We are not waiting for safety to arrive. We are building it in how we protect each other, uplift each other, and allow each other to be seen.

So, what if visibility were safety?

Not the exception, but the expectation. Not a gamble, but a guarantee. Not the edge of danger, but the beginning of freedom.

Chapter 4: Hood Epigenetics

Hood Epigenetics— Trauma, Memory, and Molecular Legacy

"Pain does not go away. It has to be dealt with, or else it will be passed on. It does not die with the person. It is transmitted." -Toni Morrison.

In the chapters leading to this one, we explored the ways Black and Brown men navigate a web of internalized and externalized oppression. Masculinity emerged not as a fixed identity, but as a risk-laden performance, one that discourages vulnerability, rewards dominance, and punishes help-seeking. We examined how strength becomes armor, how silence becomes discipline, and how survival often requires emotional contraction. We also journeyed through the politics of visibility, particularly for LGBTQ+ Black and Brwon men, whose daily lives demand constant calibration, when to pass, when to perform, when to disappear. From barbershops to churches, from street corners to home kitchens, we witnessed how exclusion, ridicule, and hyper-surveillance fragment the self. The labor of managing masculinity, sexuality, and safety is relentless. But the pressures described in earlier chapters are not merely psychological. They are biochemical.

This chapter is where all those threads converge. Here, we move from the social and emotional terrains for Black and Brown manhood into the biological realm, examining how chronic stress and structural violence do not simply shape experience, but alter the body itself. Trauma does not travel across generations only through stories, scars, and behaviors. It travels through molecules and memory. What happens to a body under siege when it becomes a bloodline?

There are wounds that do not scar the skin but live in the blood. Trauma, when unaddressed doesn't just linger in memory, it embeds itself in physiology. It alters how genes are expressed, reshapes stress-response systems and quietly maps itself onto future generations. For Black and Brown men living in urban environments, prolonged exposure to racism, poverty, violence, and environmental degradation does not only increase rates of chronic stress or depression, it accelerates biological wear. From shortened telomeres to dysregulated cortisol cycles, the streets tell a molecular story. Epigenetic research now confirms what our communities have long known intuitively: the hood leaves fingerprints on our DNA. These are not metaphors. They are methylation patterns. The legacy of trauma is not just emotional, it's inheritable.

The Legacy of Trauma
Before a child draws their first breath in cities like Baltimore, St. Louis or Charleston, the legacy of trauma is already present. This legacy is not abstract nor is it merely historical. It is rooted in centuries of structural racism, slavery, Jim Crow segregation, redlining, police violence, mass incarceration, and medical neglect. For Black and Brown communities, especially men, these systemic assaults have never been just historical footnotes, they are active, lived experiences. The biological consequences of this history are now visible in scientific research. Patterns once understood only through lived experience such as chronic stress, early illness, premature deaths are

increasingly recognized as manifestations of embodied inequality. Public health scholars have long argued that racism operates as a chronic exposure rather than a discrete event, producing cumulative physiological wear and tear across the life course.[1] Trauma, in this sense has a lineage. It is passed not only through stories and silences but through biological instructions that shape how the body responds to stress, threat, and care.

Slavery did not end so much as it evolved. Its logic reappeared in Jim Crow segregation, mass incarceration, discriminatory housing, labor exploitation, and the criminal legal system. Each iteration reproduced conditions of surveillance, deprivation, and exclusion. Research confirms that the persistent exposure to racism is associated with accelerated biological aging and dysregulated stress response systems. A 2018 study published in the Proceedings of the National Academy of Sciences (*PNAS*) found that African American individuals who reported higher levels of perceived discrimination showed markers consistent with accelerated aging, including shortened telomeres and heightened physiological stress burden.[2] This is not coincidental. These findings reflect what Geronimus termed "weathering": the idea that inequality becomes biologically embedded through repeated adaptation to adversity.[3]

Within families, trauma is often transmitted quietly. Fathers, shaped by their own generations of racialized expectations of stoicism and emotional control may pass down silence more readily than softness This inheritance is solely emotional. Studies of intergenerational trauma demonstrate that children of trauma survivors can exhibit altered stress physiology such as elevated cortisol levels, heightened threat sensitivity even in the absence of direct exposure to violence.[4] What is learned as survival becomes instruction. This intergenerational transmission is not always visible, but it is palpable. It's in the lowered gaze, the clenched jaw, the unspoken grief around dinner tables. It's in who gets to cry, and who must be "a man."

The paradox of resilience sits at the center of this legacy. Strength, endurance, and emotional control have functioned as cultural anchors in the face of systemic harm. Yet these same adaptations often foreclose opportunities for healing. Communities survive through grit, but bodies remember the cost of that survival. The language of pride—*we made a way out of no way*—exist alongside unspoken grief, suppressed vulnerability, and cumulative strain.

To understand the health of Black and Brown men in urban environments, we must begin with this legacy, not as a metaphor but as a mechanism. The trauma of the past is not buried; it is expressed in the present. It echoes in cortisol spikes, in heart disease statistics, in premature deaths, in a boy's quiet flinch when someone raises their voice. These aren't just outcomes. They are symptoms of a long, deliberate social pathology. And as we will see in the next section, science is beginning to catch up with what many have known in their bones for generations.

Epigenetics 101

There is a long-held belief that deoxyribonucleic acid (DNA) is destiny; that the genes we inherit from our parents determine who we will become and what illnesses we will face. Epigenetics tells a more complicated and empowering story. Rather than focusing on the genes themselves, epigenetics examines how our environment and experiences shape *how* those genes are active and which stay dormant. A helpful metaphor is music. Our genetic code can be imagined like sheet music (fixed and unchanged). Epigenetics is the conductor, deciding which instruments/sections of the orchestra are played and which stay silent altogether. And just like a conductor can change the mood and tempo of a song, our lived experiences like joy, trauma, safety, stress, love can shape how our biology sounds.

One of the primary biological mechanisms of epigenetics is DNA methylation. This process involves small chemical tags, called methyl groups that attach to our DNA and either turn genes on or off. In the simplest terms, methylation is the body's way of placing tiny chemical notes on genes that say, "read this more," "read this less," or "don't read this at all." These changes don't alter the genetic code itself, but they do change how the code is read. When a gene is heavily methylated, it is often silenced. When methylation is reduced, the gene may become more active.

Importantly, methylation is responsive. It changes in response to lived experience. Scientists have found that childhood trauma, poverty, exposure to violence or even being raised in a neighborhood with constant police presence can leave epigenetic "scars" on genes involved in stress response and immune signaling.[5] These are not temporary shifts; they can stay with a person for life, and in some cases, be passed down to their children. This idea might sound abstract, but the evidence is concrete.

A widely cited study of Holocaust survivors and their children found that both generations had altered methylation in genes tied to stress regulation.[6] That means the children. Who hadn't lived through the trauma, still carried a molecule imprint of it. Parallel patterns have been observed in populations exposed to racialized and structural trauma. Research examining Black Americans living in the long aftermath of slavery, Jim Crow, redlining, and mass incarceration suggest that historical and contemporary racism shapes biological regulation in ways that mirror other forms of collective trauma.[7] These findings underscore a crucial point: the past does not disappear. It becomes embodied.

Epigenetics explains how experience becomes biology, but it does not yet tell us what that conversion cost over time. To understand the full weight of this molecular memory, we must look at how chronic stress reshapes the body's internal clock. The question is no longer whether trauma gets under the skin, but how long the body can carry it before it begins to age differently. Cortisol cycles, cellular repair systems and telomeres, the protective caps that mark biological time that offer a window into how inequality is not just endured but measured. In the next section, we turn to the biology of accelerated aging where stress does not simply register as memory, but as lost years.

Telomeres, Cortisol, and Accelerated Aging

If epigenetics explains how experience writes itself into the genome, then cortisol and telomeres reveal how long the body has been forced to endure. Stress does not merely activate survival response; it keeps the time. It measures how often the alarm is sounded, how rarely the body is allowed to rest, and how quickly the biological clock is pushed forward. For Black and Brown men navigating urban life under constant vigilance, stress becomes not a momentary state, but a permanent condition.

A central biological pathway in this process is cortisol, the body's primary stress hormone. Cortisol is essential for survival. In moments of danger, it mobilizes energy, sharpens attention, and helps the body respond. But when stress becomes chronic, as it often does under conditions of racial surveillance, eviction notices, food insecurity or surveillance, cortisol stays elevated. Over time, this dysregulation alters gene expression in ways that impair emotional regulation, increase vulnerability to disease, and disrupt the body's ability to return to baseline after stress.[8] For Black and Brown men in urban spaces, this means that navigating daily life isn't just emotionally exhausting, it becomes biologically toxic.

Epigenetics also helps explain how chronic stress accelerates biological aging through its effect on telomeres, the protective caps at the ends of chromosomes. Telomeres are protective caps on the ends of chromosomes, like the plastic tips on shoelaces that keep them from fraying during cell division. Each time a cell divides, telomeres naturally get a little shorter. On average, about 20 to 40 base pairs of telomeric DNA are lost with every cell division.[9] When they become too short, cells age and die. But chronic psychological stress, trauma and environmental hardship have been linked to accelerated telomere shortening, essentially speeding up the body's biological clock.[10] Multiple studies have found that Black men, particularly those exposed to lifelong racism or sustained socioeconomic stress, have been shown to have shorter telomeres than their white counterparts, even after adjusting for chronological age.[11] This phenomenon isn't just aging, it is weathering, the cumulative biological toll of living under persistent structural strain.

Research has shown that neighborhood stressors, like living in areas with high levels of violence, poverty and environmental degradation can significantly shorten telomeres. A study conducted by the University of California, San Francisco, found that individuals living in high-stress urban neighborhoods had telomeres that were on average, 7-8 years shorter than those living in lower stress.[10] For Black and Brown men, this means that the daily stresses of urban environments, be it fear of police violence, exposure to crime or living in food deserts could literally be speeding up their biological aging.

The impact of these environmental stressors is even more striking when we consider that Black men are disproportionately affected by chronic stress. Research has shown that Black men, especially those living in lower-income neighborhoods, tend to have significantly shorter telomeres that their white counterparts. One study published in *Social Science & Medicine* found that Black men had 9-10% shorter telomeres than white men, even after adjusting for age and other variables.[11] This difference is a reflection of the accumulated burden of living in environments filled with racialized violence, poverty and exclusion. Importantly, epigenetics doesn't only explain harm, it also opens the door to healing. If trauma and stress can alter gene expression, then safety, community, love, and cultural affirmation may be able to reverse some of that damage. Emerging research shows that supportive environments, mindfulness practices, stable housing, and even social connectedness can lead to positive epigenetic changes.[12] The same biology that carries the memory of trauma can also remember healing. As we move forward, the question is no longer whether stress gets under the skin, the evidence is clear that it does. The deeper question is what happens next: how these biological imprints are passed forward, and how cycles of stress become intergenerational.

That story unfolds in the next section.

Inherited Burdens—What We Pass On
Trauma does not end with the individual. A growing body of research demonstrates that chronic and systemic stress leaves biological imprints that can be transmitted across generations. This process, often described as transgenerational epigenetic inheritance, suggests that environmental exposures and psychosocial stressors experienced by parents can alter the biological blueprint they pass on to their offspring. One of the most widely cited bodies of evidence comes from studies of Holocaust survivors and their children. Yehuda et al. (2016) found that children of Holocaust survivors exhibited altered cortisol regulation and changes in the *FKBP5* gene methylation, even though they had never directly experienced the trauma themselves.[6] Similar findings have emerged from populations affected by the Rwandan genocide, apartheid in South Africa, and residential schooling in Indigenous communities, each demonstrating that trauma embeds itself not only in culture and memory, but in biology.

Among Black Americans, the legacy of slavery, segregation, forced migration, redlining, and mass incarceration has created a unique and persistent trauma pipeline. Studies like those by Kuzawa & Sweet (2009) show that chronic stressors rooted in racism and inequality contribute to poor birth outcomes, elevated blood pressure, and early onset of disease, not only in those directly impacted, but also in their children and grandchildren.[7] When fathers live in neighborhoods marked by concentrated poverty, community violence, and environmental degradation, these exposures can affect their reproductive biology, potentially influencing the health of future offspring.

Importantly, intergenerational transmission is not limited to maternal pathways. Emerging epigenetic research suggests that paternal stress exposure can also influence offspring biology. Chronic stress alters hormonal regulation, inflammatory signaling, and gene expression in ways that may shape what fathers pass on, not only through behavior, but through biology. While the precise mechanism continues to be studied, the evidence increasingly suggests that paternal bodies carry environmental memory that can influence the health trajectories of future generations.

Alongside biological transmission, trauma is also inherited through emotional and relational scripts. In many Black communities, survival has depended on emotional suppression, stoicism and hypermasculine posturing as protection against a hostile world. These strategies have enabled endurance, but they also constrain vulnerability and emotional expression. Fathers shaped by racialized pressure to appear "strong" may pass on silence more readily than softness. This inheritance is both behavioral and physiological.

Research has shown that men who chronically suppress emotions have higher levels of cortisol, a stress hormone. This constant state of hyperarousal can cause chronic health problems, including cardiovascular disease, immune dysfunction, and increased vulnerability to psychological disorders.[13] One study found that men who emotionally suppress have 38% higher cortisol levels in the evening, a significant marker of stress.[14] These patterns of stress, passed from father to child, create not only psychological but biological inheritance, especially in how their offspring's stress response develops. Their children, exposed to these inherited patterns of emotional suppression and chronic stress, are at an elevated risk of developing stress-related disorders including PTSD and anxiety, before reaching adulthood.[15]

These patterns are not accidental. They are reinforced by historical policies that actively disrupted Black family structures, including welfare regulations that penalized the presence of Black and Brown men in the household during the mid-20th century.[16] Such policies not only fractured families but reinforced emotional distance and constrained fatherhood roles, further shaping the emotional inheritance passed down through generations. The result is a layered transmission: biological stress regulation shaped by environment, emotional norms shaped by survival, and historical policies that intensified both. Trauma, in this sense, does not simply repeat—it reorganizes. It teaches bodies how to respond before danger arrives. It scripts stress responses before language forms. And unless interrupted, it quietly instructs the next generation on how to live inside a world already marked as unsafe.

Interrupting the Cycle— Healing as Resistance

If trauma can be inherited, so can healing. Just as chronic stress and adversity leave molecular imprints on the body, emerging research suggests that safety, connection, and emotional regulation can also shape gene expression in protective ways. In this sense, healing becomes not only personal or communal; it is biochemical. It becomes resistance enacted at the level of the cell.

Epigenetics remind us that biology is responsive. The same mechanisms that allow stress to dysregulate cortisol, shorten telomeres, and alter gene expression also make the body receptive to repair. Studies increasingly show that interventions which reduce stress and increase emotional regulation can influence molecular pathways associated with inflammation, aging, and stress response. Healing does not erase history, but it can interrupt how that history is carried forward.

Mindfulness practices, long rooted in Indigenous and Eastern traditions and increasingly studied within Western biomedical frameworks offer one example of this interruption. A 2014 meta-analysis by Kaliman et al. found that just one day of intensive mindfulness mediation reduced the expression of genes related to inflammation, including *RIPK2* and *COX2*, in experienced mediators.[17] Other studies have linked meditation and yoga to telomerase activity, the enzyme that protects and maintains telomere length.[18] These findings do not suggest that meditation "cures" trauma, but they do demonstrate that the biological consequences of stress are not fixed.

For Black and Brown men, healing must also be understood within a cultural and structural context. Access to traditional mental health services has long been obstructed by stigma, cost, racial bias in care, systemic underfunding, and historical mistrust of medical institutions. In response, culturally responsive forms of healing have emerged that reframe vulnerability not as weakness, but as survival. Black therapist, healing circles, peer-led support groups, and community-based mental health spaces challenge the inherited mandate of stoicism by restoring emotional literacy as a form of strength. These spaces do more than offer coping strategies; they create conditions of safety where the nervous system can finally stand down.

Healing is most powerful when it is culturally relevant. In many Black and Brown communities, wellness cannot be imposed from the outside; it must be reclaimed from within. Barbershops, for instance, are not just grooming spaces but vital forums for health education, mental health dialogue, and political discourse. Initiatives like *The Confess Project*, which trains barbers as mental health advocates for young Black men, have demonstrated significant impact in reducing stigma and improving wellbeing.[19] Similarly, faith-based spaces, while often conservative spaces, have also provided platforms for trauma-informed care. Faith-based interventions in cities like Baltimore have integrated spiritual counseling with behavioral health services, offering a culturally grounded avenue for men to process grief, anxiety, and rage. Storytelling traditions, whether through spoken words, memoirs, or intergenerational dialogue, allow pain to be transformed into narrative, and narrative into power. This oral transmission of struggle and survival becomes a bridge toward collective healing.

In Detroit, the *Black Men's Mental Health Project* uses a combination of group therapy, physical activity, and financial literacy coaching to address the multiple layers of stress affecting men's in the city. Preliminary evaluations have shown improvements in participants' self-efficacy, reduced depressive symptoms, and stronger community bonds. In Baltimore, *Healing City Baltimore* emerged in response to youth trauma and neighborhood violence. It convenes residents, city officials, and health professionals to implement trauma-informed care across schools, social services, and law enforcements. Its centerpiece, the Elijah Cummings Healing City Act, mandates trauma-responsive training for all city agencies, institutionalizing healing as a public health strategy.

In Oakland, the *Roots Community Health Center* combines culturally competent clinical care with reentry support, job training, and mental health services. Their "Emancipator Initiative" targets men impacted by incarceration and poverty, recognizing that reentry without healing only perpetuates the cycle of trauma. Their integrated model has been linked to reduced recidivism and improved chronic disease outcomes.[20] These efforts are not merely interventions; they are acts of defiance against a system that has normalized Black and Brown male suffering. To heal in a society that profits from your pain is to resist. And to organize that healing collectively is to rewire the molecular and social legacy we pass on. When Black men are given the space to feel, to process and to heal, the cycle is not just paused, it is disrupted. In the next section, we turn to the science of resilience and ask how the body, when nurtured, can reprogram itself toward a future no longer haunted by inherited trauma.

Rewriting the Molecular Narrative

While trauma can leave molecular scars, it does not write the final chapter. Epigenetics is not destiny, it is potential. The same biological mechanisms that allow stress to mark the genome also allow healing to modify it. In other words, just as methylation patterns can silence or activate genes in response to toxic stress, they can also respond to care, love, and safety. This flexibility, what scientists refer to as epigenetic plasticity, means the molecular narrative is not fixed. It can be edited, revised, and in some cases, fundamentally rewritten.[21] Emerging research in neuroplasticity and epigenetic remodeling suggests that positive experiences and sustained psychosocial support can recalibrate stress-response systems that have been dysregulated by trauma. Supportive relationships, stable housing, mindfulness practices, and culturally affirming social bonds have been associated with improved regulation of cortisol, reduced inflammatory signaling, and shifts in gene expression linked to emotional regulation.[22] These changes do not erase history, but they demonstrate that the body remains responsive to care long after harm has occurred.

Rewriting the molecular narrative is not a biological process; it is also a narrative one. Black and Brown men often inherit not only trauma but also silence. Breaking that silence, through therapy, group dialogues, barbershop conversations, oral storytelling, or spoken word offers an opportunity to reframe the past. When men begin to narrate their own experiences of pain, survival, and transformation, they are doing more than self-expression. They are restructuring neural pathways and reshaping stress response systems.[23] Words, in this sense, are both symbolic and biological tools of recovery.

This rewriting also occurs in fatherhood. When a Black father chooses to be emotionally present despite having no model for vulnerability, he interrupts a cycle. When he tells his son, "You don't have to be strong all the time," he is not only offering emotional permission, he is potentially altering how that child's body process stress. Studies show that parental warmth and emotional support can buffer the physiological impacts of stress and even influence methylation patterns in genes linked to emotion regulation.[24] These micro-acts of emotional generosity are molecular revolutions. They are the invisible but powerful work of repair happening in kitchens, barbershops, classrooms, and community centers.

Importantly, reframing the molecular narrative is not about romanticizing resilience or placing the burden of healing solely on individuals to heal themselves in unjust conditions. It is about reclaiming agency within constraint. In a world that systematically denies Black and Brown men safety, softness, and longevity, choosing to heal to feel, to connect, to tell the truth becomes an act of resistance. It asserts that biology is not only a record of harm, but a site of possibility. In the end, the science of epigenetics gives us language for something our ancestors always knew that healing is generational, and that it matters how we live, love, and relate to one another. The molecular narrative, once assumed to be inherited and immutable, is instead a living manuscript. One we are still writing.

Yet rewriting the body is only part of the work. Healing at the molecular level cannot be sustained without transforming the environments that etched trauma into the genome in the first place. The chapters that follow, move from memory to material conditions. Chapter 5 turns outward, examining how heat, housing, and hormonal disruption converge in urban environments, and how climate, infrastructure, and inequality become biological insults that shape health long before illness appears.

Chapter 5: Heat, Hormones, and Housing

Heat, Hormones, and Housing—Environmental Endocrinology in the Inner City

"The body keeps the score, but the environment writes the script."- Bessel van de Kolk.

In urban environments, particularly in marginalized neighborhoods predominantly composed of Black and Brown residents, the intersection of environmental stressors, economic precarity, and health disparities is stark. The phenomenon of urban heat islands (UHI) is a striking example of how environmental inequality can shape the lived experiences of these communities. For Black men, already navigating a host of social determinants of health, such as systemic racism, limited access to healthcare, and economic instability, the added burden of extreme heat compounds their vulnerabilities. The UHI effect in low-income neighborhoods exacerbates the physiological and psychological toll on the body, resulting in heightened levels of stress, hormonal imbalances, and a greater risk for chronic health conditions. This section explores how the urban heat island effect, fueled by historical and ongoing environmental inequities, impacts Black and Brwon men, drawing connections between climate and health disparities while addressing the profound implications on their endocrine systems.

Heat Islands and Hormonal Havoc

Urban heat is not just a seasonal inconvenience; it's a public health emergency unfolding in slow motion. In the dense concrete-bound neighborhoods of many American cities, heat doesn't dissipate. It accumulates, radiates, and lingers. This phenomenon, known as the urban heat island (UHI) effect, occurs when neighborhoods dominated by asphalt, concrete, and dense infrastructure absorb and retain heat while lacking the green space necessary for cooling.

The burden of this heat is not randomly distributed. It is racially and economically patterned. Historically redlined Black and Brown neighborhoods are up to 7°F hotter than nearby wealthier, greener and predominantly White neighborhood areas nearby.[1] These same communities have 30%-52% less tree canopy and nearly double the proportion of heat-retaining surfaces like asphalt and concrete.[2] These same neighborhoods are also less likely to have access to air conditioning, energy-efficient housing, or cooling centers, amplifying the risk of extreme heat exposure. In these environments, heat does not simply arrive, it lingers, day and night, turning homes, sidewalks, and streets into slow cookers of stress and physiological strain.

This isn't just about being uncomfortable or sweaty. Prolonged heat exposure to elevated temperatures places sustained stress on the endocrine system, which regulates hormones. One of the key players in the hypothalamic-pituitary-adrenal (HPA) axis, the body's central stress response network. You can think of it as a built-in thermostat for survival: when your body senses heat as a threat, the brain sends signals to the adrenal glands to release cortisol, the main stress hormone. Occasional activation of this system is adaptive. Chronic activation is not. Repeated heat exposure without adequate recovery disrupts normal cortisol rhythms, contributing to sleep disturbance, chronic fatigue, immune suppression, and emotional dysregulation.[3] What Chapter 4 describes as a biological mechanism becomes, here, an environmental trigger—one that fires relentlessly during long summers and increasingly frequent heat waves.

For men, this heat-induced endocrine disruption extends beyond cortisol. Studies show that high temperatures are associated with lower testosterone levels, reduced sperm count, and abnormal sperm morphology, potentially impacting fertility and long-term reproductive health by as much as 12%.[4] These hormonal shifts intersect with heightened risks of cardiovascular disease, type 2 diabetes, and mental health disorders conditions already disproportionately affecting Black men in urban settings.[5] What seems like "just heat" is actually reshaping the body's hormonal architecture.

The uneven geography of heat mirrors the geography of disease. In cities such as Baltimore, Detroit, and Los Angeles, neighborhoods with the highest surface temperatures are also those with the lowest median incomes and highest chronic disease rates.[6] For the men living in these environments, enduring summer heat without reliable air condition, tree cover, or safe cooling spaces becomes a daily act of invisible labor, one that is hormonally taxing and physiologically unstainable. Heat, in this context, is more than a weather phenomenon; it's a form of embodied inequality, with cascading consequences for masculinity, biology and urban survival.

The physical environment, where heat, poor housing, and environmental toxins intersect, creates an even greater challenge for residents, particularly in communities that already bear the burden of systemic inequities. In these urban spaces, where the effects of the urban heat island (UHI) phenomenon are compounded by inadequate housing, the risk of environmental stress is magnified. Poor housing conditions, which often lack the necessary insulation, cooling infrastructure, and air quality controls, contribute to a dangerous feedback loop that further disrupts the body's natural systems. In the next section, we turn indoors to examine how inadequate housing conditions in vulnerable neighborhoods can amplify the endocrine system, setting the stage for long-term health challenges that are especially pronounced for men in marginalized communities.

Hormone-Disrupting Housing

Housing is more than just a place of residence; it is a powerful determinant of health, shaping not only the physical environment but also the very systems that regulate the body. In urban areas burdened by extreme heat, poor housing quality function as an amplifier, intensifying thermal exposure, trapping pollutants, and sustaining chronic physiological stress. For Black and Brown men living in historically disinvested areas, housing conditions often convert environmental heat into a constant endocrine assault.

Much of the urban housing stock in marginalized communities is older, poorly insulated, and ill-equipped to manage rising temperatures. Homes built before modern energy standards frequently lack adequate ventilation, reflective roofing, or central air conditioning. As a result, indoor temperatures can exceed outdoor temperatures during heat waves, especially in top-floor apartments and row homes constructed with heat retaining materials. This creates prolong exposure to elevated temperatures that prevents physiological recovery, even overnight. The absence of thermal relief disrupts circadian rhythms and sustains activation of the hypothalamic-pituitary-adrenal (HPA) axis, reinforcing cortisol dysregulation and sleep disturbance.

Housing inequality also determines exposure to hormone-disrupting chemicals. Older urban homes are more likely to contain lead-based paint, asbestos, and synthetic building materials that emit volatile and semi-volatile organic compounds (VOCs and SVOCs).

Approximately 29 million of low-income homes built before 1978 still contain lead-based paint hazards, and 35% of public housing units are estimated to have asbestos-containing materials (EPA, 2015).[7] Indoor air sampling studies show that concentrations of VOCs can be 2-5x higher indoors than outdoors, particularly in poorly ventilated homes.[8]

These exposures matter because many of these compounds act as endocrine disruptors. Flame retardants, phthalates, and other SVOCs interfere with hormonal signaling by mimicking or blocking endogenous hormones, including cortisol and testosterone. When combined with chronic heat exposure, these toxins increase physiological stress load rather than functioning as isolated risks. The endocrine system does not experience heat, pollution, or housing instability separately; it integrates them simultaneously.

Access to air conditioning further illustrates how housing becomes a hormonal determinant. According to the American Housing Survey (2019), 16% of Black households and 18% of Latino households in the U.S. do not have access to air conditioning, compared to just 8% of White households.[9] For men in these environments, the inability to cool the body after heat exposure sustains elevated cortisol levels and impairs nighttime recovery. Sleep fragmentation caused by excessive indoor heat has been linked to insulin resistance, blood pressure dysregulation, and suppressed testosterone production.[10] All outcomes that are already unequally distributed along racial and socioeconomic lines.

Housing instability compounds these effects. Overcrowding, frequent relocation, and threat of eviction intensify psychological stress while limiting control over environmental conditions. A report by the National Low Income Housing Coalition (2020) revealed that for every 100 low-income renters, only 37 affordable and available rental homes exist.[11] This shortage often forces marginalized families into substandard housing where heat exposure, toxic materials, and poor ventilation are unavoidable. Chronic uncertainty around housing reinforces HPA axis activation, transforming environmental stress into a persistent biological state. For Black and Brown men, the physiological consequences of hormone-disrupting housing extend beyond discomfort. Chronic cortisol elevation, sleep disruption, and reduced testosterone are associated with fatigue, mood dysregulation, metabolic dysfunction, and increased cardiovascular risk.[12] These outcomes are often framed as individual health failures, yet they emerge from structural conditions that constrain environmental control.

Under these circumstances, housing is not simply shelter, it is an endocrine environment. Poor-quality housing does not just expose residents to heat; it converts heat into long-term hormonal stress. Understanding housing as a biological stressor clarifies why environmental health disparities persist even when individual behaviors remain constant. In the next section, we move from built environment to temporal exposure, examining how climate stress functions as chronic stress across seasons and years.

Climate Stress as Chronic Stress

Climate change is not just a matter of fluctuating temperatures; it represents a significant environmental stressor that can disrupt the body's physiological processes, particularly when experienced over prolonged periods. Chronic exposure to environmental stressors such as extreme heat, air pollution, and housing instability creates a continuous stress response in the body, leading to long-term physical and mental health consequences. These stressors do not just threaten the immediate well-being of individuals; they trigger chronic activation of stress systems that can enduring effect on hormonal regulation, particularly the secretion of cortisol, adrenaline, and norepinephrine.

Adrenaline (epinephrine) and norepinephrine, hormones released by the adrenal glands, are central to the body's fight or flight response, increasing heart rate, blood pressure, and alertness during acute stress. While these hormones are crucial in helping individuals respond to immediate threats, chronic secretion, caused by repeated exposure to environmental stressors such as excessive heat and poor air quality, can have detrimental long-term effects. For men, especially those in marginalized communities who are subjected to compounded stressors, chronic activation of the sympathetic nervous systems (SNS), which governs the release of these hormones, can contribute to cardiovascular issues such as hypertension, arrhythmias, and increased risk of heart attacks.[13] Additionally, long term secretion of adrenaline and norepinephrine has been linked to elevated levels of anxiety, irritability, and aggression, which can impair overall mental well-being.[14]

When adrenaline and norepinephrine are consistently released due to chronic stress from environmental stressors, such as extreme heat in urban areas, vascular dysfunction becomes a concern. Chronic exposure to these hormones can narrow blood vessels, increase the likelihood of blood clot formation, and promote the development of atherosclerosis (the thickening and hardening of artery walls due to the buildup of fatty materials).[15] This exacerbates the risk of cardiovascular disease, a condition that disproportionately affects marginalized populations.

The effects of chronic stress are particularly significant for Black and Brown men living in low-income urban communities. These individuals not only face physical stressors like excessive heat and poor air quality but also endure socioeconomic challenges such as housing instability, poverty, and racial discrimination factors that compound stress and increase vulnerability to mental health and physical health problems. For example, research by Cohen et al. (2020) has shown that Black and Brown men in urban areas often experience higher levels of chronic stress and mental health challenges, including anxiety, depression and post-traumatic stress disorder (PTSD), due to the combined effects of environmental pollution and systemic inequality.[16] Chronic stress from these sources is linked to lower levels of testosterone, which can lead to mood disorders, decreased energy levels, diminished sexual health and lower overall life satisfaction.[17]

Moreover, housing instability, a pervasive issue in many marginalized neighborhoods, further amplifies the physical and mental health impacts of environmental stress. Many homes in these areas are poorly insulated and lack proper ventilation, exacerbating exposure to heat, volatile organic compounds (VOCs), and particulate matter. This combination of environmental and social stressors triggers the release of cortisol and other stress hormones in the body. Over time, this can lead to cortisol dysregulation, which disrupts immune function, metabolic processes, and sleep patterns.[18] Chronic disruptions in these physiological systems can lead to long-term health consequences, including obesity, diabetes, cardiovascular disease, and mental health disorders, conditions that are already prevalent among Black and Brown men in low-income urban neighborhoods.

In sum, the chronic stressors induced by environmental factors such as heat, pollution, and poor housing conditions create a relentless strain of the body, particularly in marginalized communities. This ongoing stress response, manifested through hormonal dysregulation and heighten vulnerability to mental and physical health issues sets the stage for a cascade of negative outcomes that disproportionately affect Black and Brown men. The intersection of environmental stress and social determinants of health deepens health disparities, influencing everything from cardiovascular health to mental well-being. The compound effects of environmental stress and societal expectations of masculinity converge in ways that amplify risks, influencing men's behavior, mental health, and physical well-being. In environments where heat, societal pressures, and systemic inequality intersect, gender norms intensify vulnerabilities, making men more susceptible to aggression, violence, and risky behaviors. This complex interplay underscores the urgent need to examine the relationship between environmental stressors and masculinity, where the consequences are far-reaching, impacting not just individual health but entire communities.

Masculinity Under Heat— Hormones, Behavior, and Risk

Extreme heat does not only tax the body; it reshapes behavior. In urban environments where high temperatures persist without adequate relief, physiological stress intersects with social expectation, particularly those surrounding masculinity. For men living in marginalized neighborhoods, heat exposure compounds existing pressure to perform strength, emotional control, and endurance under conditions that actively undermine bodily regulation. The result is a volatile feedback loop in which hormonal stress responses interact with gender norms, shaping behavior in ways that are often misread as individual pathology rather than environmental consequence.

From a biological standpoint, prolong heat exposure elevates cortisol and catecholamines such as adrenaline and norepinephrine—hormones associated with vigilance, reactivity, and threat response. As discussed in earlier sections, chronic activation of these systems impairs emotional regulation, disrupts sleep, and depletes metabolic reserves. When these physiological changes occur alongside testosterone suppression, a pattern observed under sustained heat stress, men may experience irritability, fatigue, reduced impulse control, and emotional dysregulation.[19] These shifts do not occur in isolation; they are filtered through cultural scripts that dictate how distress is expressed.

Masculinity norms in many urban contexts discourage vulnerability and emotional disclosure, rewarding stoicism and dominance instead. Under heat stress, when the body's capacity for self-regulation is already compromised, these norms can transform internal distress into outward aggression or risk-taking. Research has shown that men are more likely to externalize their emotional responses, especially when they are under pressure to conform to the social expectations of masculinity, which value strength and emotional control.[20] In high stress environments, such as those that occur during extreme heat events, men may experience difficulty managing their emotions, leading to aggressive outbursts or violent behavior. These compounded pressures, both physiological and cultural, can lead to an increased likelihood of aggressive responses when dealing with the frustrations that arise from environmental heat.[21] Empirical studies consistently show a relationship between ambient temperature and violent behavior. Anderson (2001) found that violent crimes increase during periods of sustained heat, while later analyses estimate that assaults and interpersonal violence rise between 5-12% during heatwaves.[22] These increases are more pronounced in low-income urban neighborhoods where exposure is highest and coping resources are lowest.

Importantly, these behavioral outcomes are not simply products of "hot tempers" but of constrained choices. Heat-related dehydration, sleep deprivation, and endocrine dysregulation impair executive functioning, narrowing the window for adaptive decision making.[23] In this context, aggression may function as a maladaptive coping strategy, a socially legible response to overwhelm in environments where emotional expression is penalized and retreat is unavailable. For Black and Brown men, these responses are further shaped by racialized surveillance and criminalization, increasing the likelihood that stress-induced behaviors are met with punishment rather than care.

The psychological toll of heat stress compounds this dynamic. Research by Dunn et al. (2020) has shown that prolonged exposure to high temperatures can significantly disrupt cognitive and emotional processing, leading to greater emotional volatility.[23] This volatility, in turn, heightens the likelihood of aggressive behavior as individuals may have difficulty regulating their responses to frustration. One key study by Cohn et al. (2018) found that the effects of heat on aggression are particularly pronounced in communities that have been historically marginalized, where people are more likely to live in inadequate housing and face additional stressors related to poverty and systemic inequality.[24] In these environments, individuals may have fewer resources to cope with heat, leading to a greater vulnerability to the stressors associated with it. These stressors can be compounded by the daily pressures of living in marginalized communities, resulting in increased aggression and violent behavior. For instance, during extreme heatwaves, aggression levels in these areas were reported to rise by 9-15% (Zhao et al., 2019), pointing to the critical intersection of heat exposure and urban inequality.[25] These stressors are often invisible, masked by cultural narratives that equate masculinity with emotional containment. The reality is far more complex: emotional suppression under extreme conditions often gives rise to depression masked as aggression of hyperactivity, a pattern particularly dangerous for young men of color who already face criminalization in public spaces.

Moreover, the increased aggression linked to heat stress can have serious public health implications, especially for marginalized men. Research has shown that aggressive behavior in response to heat stress not only increases interpersonal violence but also intensifies mental health challenges. For instance, a study by Zavaleta et al. (2015) indicated that during periods of high heat, men in low-income urban communities are more likely to engage in behaviors that contribute to social unrest including violent confrontations and risky decision-making.[26] The combination of heat induced stress and societal expectations of masculinity creates a dangerous cycle in which aggression becomes an immediate coping mechanism in the face of overwhelming environmental stress. For Black and Brown men, this aggression can be exacerbated by experiences of racial discrimination and social exclusion, which further compound their vulnerability to the mental and physical health impacts of heat stress. As these men are often faced with the dual pressures of systemic inequality and environmental degradation, the likelihood of experiencing aggression or engaging in violent behavior during heatwaves increases, which can perpetuate cycles of harm within these communities.

Coping with extreme heat requires resources; air-conditioning, access to green spaces, adequate housing insulation, all of which are unevenly distributed along racial and economic lines. Marginalized men are often denied these buffers not only by circumstance but also by masculine codes that valorize endurance. This creates a paradox: to cope effectively would require vulnerability and help-seeking, yet to do so is perceived as betrayal of masculinity. The result is often maladaptive coping aggressive, withdrawal, substance use, behaviors that provide short-term relief but long-term harm. Santos et al. (2017) argue that such patterns are not failures of character but symptoms of constrained choices in high-stress environments.[21]

Men in these communities are not without resilience, but their coping strategies are frequently shaped by the intersections of masculinity, race, and resource scarcity. In a study by Zavealeta et al. (2015), men living in high-heat, low-resource neighborhoods were more likely to report coping strategies centered on risk-taking, including confrontational behavior and disengagement from community support systems.[26] While these behaviors may seem irrational, they are rational responses to an environment where survival is both physical and reputational. For men who are told that vulnerability equals weakness, acts of defiance, even violent ones may be the only socially legible expressions of emotional overwhelm.

However, community-based interventions that center gender, race, and place can help disrupt these cycles. Programs that offer heat relief, mental health education, and safe spaces for emotional expression, while also challenging rigid norms around masculinity have shown promise. For example, grassroots initiatives in Baltimore and

Oakland have integrated barbershops, youth mentorship, and mental health outreach to create culturally resonant spaces for men to unpack stress and trauma. These interventions are not simply about cooling the body but about softening the rigid edges of masculinity that make heat, aggression, and silence a toxic brew. Healing, in these communities, begins not with air conditioning, but with redefining what strength looks like under stress.

Policy, Design, and Resistance

The racialized landscape of American cities did not emerge by accident; it was shaped by decades of discriminatory policy and planning practices that continue to manifest today in environmental disparities, including disproportionate exposure to extreme heat. Redlining, highway construction, and urban renewal projects have historically funneled Black and Brown communities into heat vulnerable neighborhoods, those with fewer trees, more paved surfaces, and limited green infrastructure.[27] For example, a national study of 108 U.S cities found that formerly redlined neighborhoods are on average 4.7°F hotter than non-redlined areas.[27] In cities like Baltimore and Richmond, these differences are even more stark, reaching up to 10°F where historically disinvested neighborhoods coincide with some of the most heat-vulnerable census tracts.[28] These legacies are not just about temperature; they're about policy choices that physically inscribed environmental risk into communities of color.

Furthermore, inadequate housing codes and zoning laws exacerbate indoor heat exposure. Renters, particularly in low-income, urban areas are often subject to slum conditions where landlords are not required to provide air conditioning or insulation.[29] Without tenant protections or heat ordinances, people are left to bear the physiological and psychological burden of environmental extremes alone. In New York City, a study found that low-income households were 56% less likely to own or access air conditioning and 70% more likely to report negative health outcomes during heat waves.[30] These inequities are not natural, they are engineered.

The physical layout of cities can either exacerbate or buffer against heat, depending on how equitably design elements are distributed. The urban heat island (UHI) effect is intensified by high concentrations of impervious surfaces, like asphalt and concrete and sparse vegetation. In neighborhoods like South Central Los Angeles or East Oakland, satellite imagery reveals significantly less tree canopy than in adjacent wealthier neighborhoods.[31] This display isn't simply aesthetic; it carries measurable consequences. Increased surface temperature in these areas correlate with higher emergency room visits for heat-related illnesses, particularly among Black and Latino populations.[32]

Design decisions such as the placement of cooling centers, the funding for green infrastructure, or the creation of shade corridors in public transit zones all reflect municipal priorities. Unfortunately, these features are often implemented in gentrifying areas rather in long-neglected communities, leading to what some scholars call "climate gentrification".[33] In Miami, investments in tree planting and green space upgrades in high-ground neighborhoods inadvertently displace lower income residence by increasing property values and rents.[34] These interventions, while environmentally beneficial, reproduce the same inequities they intend to address if not paired with equitable housing and anti-displacement policies.

Despite systemic challenges, communities have not been passive recipients of environmental injustice. Across the U.S., grassroots movements have emerged to reclaim environmental agency and demand policy change. In Detroit, community groups like Eastside Environmental Council have mobilized to secure funding for tree-planting initiatives and push for heat mitigation efforts that reflect local needs rather than external agendas.[35] Their organizing led to the development of the city's "Resilient Neighborhoods" framework, which explicating prioritizes historically marginalized neighborhoods in climate adaptation planning. Likewise, in Phoenix, Arizona, one of the hottest cities in North America, residents of the Maryvale neighborhood, predominantly Latinx and working class, partnered with researchers to conduct participatory heat mapping. The data revealed that their community regularly experienced a surface temperature of 15°F hotter than affluent areas like Arcadia.[36] This locally driven data collection became instrumental in lobbying city officials to reroute funding toward equitable heat relief measures, including shaded bus stops and heat health outreach in Spanish.

At the policy level, some cities are beginning to embed heat equity into their planning framework. Portland and Los Angeles now include urban heat mitigation as a public health priority in their climate action plans, and Baltimore has initiated the "Cooling Cities" initiative that targets heat interventions in neighborhoods with the highest vulnerability score.[37] However, these reforms often originate from sustained community pressure rather than top-down benevolence. As the environmental justice scholar Robert Bullard reminds us. "The people most impacted by pollution and climate change are often the ones leading the fight against it."[38]

Designing urban environments with justice in mind demands not only a technical understanding of infrastructure but a deep commitment to equity that explicitly recognizes who is most at risk. For Black and Brown men in disinvested neighborhoods, the built environment often reflects a legacy of neglect: cracked sidewalks, few trees, high-density concrete, and insufficient public spaces for cooling or rest. These physical conditions are not merely aesthetic oversights; they are racialized design choices with physiological and psychological consequences. Exposure to extreme heat compounds the daily pressures of navigating racism, economic marginalization, and hyper-surveillance. When city blocks feel both physical hot and socially hostile, Black, and Brown men may internalize this as yet another signal of disposability, further fraying mental health and increasing chronic stressors.[39]

Equitable urban design can challenge these patterns by transforming historically oppressive spaces into zones of care and resilience. For example, in cities like Baltimore and Philadelphia, community coalitions led by Black men have spearheaded projects that combine heat mitigation strategies like tree planting and cool roofing with job creation, youth mentorship and cultural reclamation. These programs demonstrate that cooling interventions don't just have to be technical, they can be healing. Initiatives like Baltimore's Cooling Cities Strategy specifically integrate workforce development and community engagement with climate resilience employing and training local residents (including Black men) to plant trees, build reflective surfaces, and install cooling systems in heat vulnerable homes.[37] By situating Black and Brown men as not only at-risk populations but also as central actors in resistance and redesign, these efforts affirm their knowledge, dignity, and stake in the urban future.

Designing for justice is not merely a matter of technical planning or aesthetic preference, it is a moral imperative that demands attention to who is protected, who is heard, and who is harmed. As we've seen through case studies and lived experiences, the consequences of policy choices, zoning laws, and neighborhood design are inscribed not only in urban landscapes, but in the bodies of those who inhabit them. For Black and Brown men, these impacts are often compounded, woven through generations of racialized neglect, economic disenfranchisement, and the expectations of masculinity that demand strength even under unlivable conditions. But what if the conditions themselves are the crisis? What if the body—its stress, its heat, its hormone levels—is telling us a political story?

To understand the full weight of environmental injustice, we must now turn inward to the biological responses shaped by systemic inequality. In this final section, we connect the dots between hormonal disruptions and structural violence. We argue that hormones are political, that what shows up in cortisol spikes, testosterone suppression, and mental health struggles is not individual failure, but collective neglect. It is here, at the intersection of biology and policy, that the fight for health equity must be waged.

Conclusion: The Hormone is Political

The body is not a neutral vessel. It carries memory, metabolizes stress, and responds to the world around it with chemical precisions. In marginalized communities, particularly among Black and Brown men, the hormonal disruptions shaped by extreme heat, chronic stress, and systemic inequality are not incidental. They are evidence of structural violence playing out under the skin. What we measure in cortisol surges, suppressed testosterone levels, and inflammatory biomarkers is not just the physiological result of hardship, but the biological imprint of political decisions.[40] These decisions about where green spaces are placed, which neighborhoods are redlined, how much public investment is allocated to heat resilience, determine not only the health of the environment, but the internal chemical environments of those who live within it.

This is why we must understand the hormone as political. It is not simply an endocrine response; it is a mirror of policy failure. For example, research shows that chronic exposure to stressors like poverty, racism, and heat correlates with elevated allostatic load that is disproportionately higher among Black men compared to other groups.[41] Elevated cortisol and dysregulated testosterone are not merely personal health issues; they are indicators of a system that continues to fail its most vulnerable. This is particularly true in urban environments where rising temperatures, inadequate infrastructure, and hyper-surveillance intersect to create a terrain where the demand to "be strong" becomes both a cultural expectation and a biological hazard.[42]

Furthermore, these hormonal shifts don't just stay inside the body. They influence how individuals respond to the world emotionally, behaviorally, and cognitively. When testosterone levels drop and cortisol remains elevated, individuals may experience fatigue, depression, reduced impulse control, or increased aggression under stress.[19] For Black and Brown men who are often policed not only by the state but by social expectations of stoicism and control, these biological effects can manifest as survival mechanisms that are easily pathologized but rarely contextualized. The public sees aggression or disengagement. Public health sees a crisis of masculinity. But what's really happening is that bodies are being shaped by systemic exposure to trauma, heat, and neglect.

To challenge these outcomes, we must go beyond individual interventions and toward policy transformation. Interventions that address hormonal dysregulation through therapeutic, environmental, and social supports must be paired with upstream changes like heat mitigation policies, equitable housing reform, and culturally grounded mental health services.[43] As long as systemic inequities remain, the hormonal response will continue to tell the truth, no matter how much we try to individualize or ignore it. Hormones are messengers not just of internal imbalance, but of external injustice. When we treat the hormone as political, we begin to ask the right questions, not just about how people cope, but about why they must.

In the end, reclaiming hormonal health as a matter of justice is not only a biological imperative, it is a political one. To heal communities, we must listen to the signals the body is sending and recognize that biology is not destiny, but a consequence of context. This means investing in healthier environments, disrupting harmful gender norms, and confronting the legacy of racist and classist policy decisions that have literally changed the body's chemistry. Health equity requires us to connect the molecular with the municipal, the hormonal with the historical. Because when we do, we reveal what has always been true: the body remembers, and it remembers who failed to protect it.

Chapter 6: Urban Immunity

Urban Immunity — Food Deserts, Inflammation, and Resistance

> "The immune system is not only a reflection of the body's internal state, but also a mirror of the society that surrounds it."-Dr. Arline Geronimus.

Urban environments are more than concrete and steel; they are metabolic landscapes that shape the immune system of those who inhabit them. For Black and Brown communities, particularly men navigating the intersecting pressures of race, poverty and masculinity, immune health becomes both a physiological and political battleground. Food deserts, food swamps, food mirages, polluted air, systemic violence, and chronic stress are not isolated hazards; they are environmental constants that chip away at immune resilience and drive persistent inflammation. Recent research shows that sustained exposure to psychosocial stressors and environmental toxins in disinvested neighborhoods elevates cortisol and inflammatory markers, increasing vulnerability to chronic diseases such as diabetes, asthma, hypertension and even certain cancers.[1] In this way, the immune system becomes an archive, documenting the body's lifelong encounter with inequality.

This chapter explores how the immune system responds to structural violence in urban settings, using food deserts and chronic inflammation as entry points into a broader story about embodied injustice. It centers the lived realities of men whose immune responses have been shaped by what they breathe, what they eat, and what they endure. Drawing on ecological immunology, critical public health frameworks, and community-level data, this chapter argues that immunity is not just biological, it is deeply spatial, racialized, and political. For marginalized men, especially those racialized as Black and Latino, access to nutritious food, clean air, and anti-inflammatory environments is not guaranteed but contested. From asthma wards in the Bronx to community gardens in Detroit, this chapter examines the geography of immune suppression and the grassroots movement pushing back. In doing so, it calls for an urgent reframing: immune health is not a matter of personal responsibility, but of collective justice.

The Geography of Hunger

The term "food desert" has long been used to describe neighborhoods, often urban, predominantly Black or Brown where fresh, affordable, and nutritious food is scarce.[2] Yet even this term fails to fully capture the systemic roots of such scarcity. Activists and scholars increasingly prefer the phrase food apartheid, which situates food insecurity with a deliberate framework of racialized disinvestment, redlining, and exclusionary zoning. Unlike the passive imagery of a "desert" food apartheid reflects how policy, race and capitalism work together to deprive communities of life-sustaining nourishment.[3] These injustices are not simply matters of geography but of design: the result of political decisions that have prioritized profit and segregation over equity.

A major contributor to this structural hunger is supermarket redlining, the practice which major grocery chains avoid or abandon low-income Black neighborhoods due to perceived financial risk. This phenomenon has been closely linked to earlier practices like redlining and racial zoning laws, which shaped where investment occurred and where it didn't.[4] The legacies are spatially enduring. In Chicago, predominantly Black neighborhoods are more than five times more likely to lack full-service grocery stores compared to white neighborhoods.[5] In Detroit, over 30% of residents live more than a mile from the nearest supermarket with limited transportation options compounding the challenge.[6]

Across many cities, what exists is not an absence but an overabundance: corner stores stocked with ultra-processed foods, fast food chains at every major intersection, and grocery stores that overcharge for lower-quality produce. These are food swamps, zones where caloric overload and nutrient deprivation co-exist, overwhelming communities with options that fuel inflammation and immune dysfunction.[7] In neighborhoods like West Baltimore or South Los Angeles, proximity to food does not guarantee nourishment. Instead, high concentrations of sugar, sodium, and trans fats saturate local diets, inflaming the body and feeding the chronic diseases such as diabetes, hypertension, cardiovascular disease that disproportionately harm Black and Brown men.[8]

In Baltimore, a city with a long history or redlining and racial segregation, the Baltimore Food Environment Map revealed that nearly one in four residents live in what the city officially designates a "Healthy Food Priority Area", a term developed to replace "food desert" that more accurately conveys systemic neglect.[9] The Baltimore Food Desert Mapping Project in particular provides a powerful case study in the visualization of racialized hunger. Using spatial data, the initiative revealed that Healthy Food Priority Areas are heavily concentrated in predominantly Black neighborhoods such as Sandtown-Winchester and Cherry Hill. These areas not only lack access to grocery stores but also suffer from high rates of poverty, lower car ownership, and greater reliance on convenience stores and fast-food outlets (food swamps). These findings underscore how structural racism, not just market dynamics, governs food availability.[9] The visualization of this data has helped frame food justice as a public health emergency, linking the geography of food to higher rates of diabetes, obesity, and hypertension in these communities.

Community voices deepen this statistical portrait. In qualitative interviews with Black residents in East and West Baltimore, many describe a daily grind shaped by constrained mobility, economic precarity, and the normalization of unhealthy options. One resident explained: "We eat what we can afford, not what we need. There's a grocery store two bus rides away, but the corner store is right there. "Another noted the psychological toll: "It's not just about hunger; it's about being forgotten." These stories often illustrate how reliance on corners stores, often stocked with ultra processed foods, become both a logistical necessity and a nutritional hazard. These accounts are echoed in cities like Detroit, where residents cite transportation barriers, food costs, and time poverty as key barriers to health eating.[6]

The health consequences are profound. A lack of access to nutrient dense foods contributes to chronic inflammation, weakens immune function, and accelerates disease such as Type 2 diabetes, hypertension, and obesity, conditions that disproportionately affect Black men.[8] In this context, poor dietary options are not a matter of individual choice but systemic imposition. The immune system, continuously under siege from processed foods and insufficient micronutrients, becomes a site where inequality becomes biological. Studies show that neighborhoods with low food access have significantly higher rates of emergency room visits for diet-related illnesses, creating a feedback loop of poor health and limited mobility.[10]

Understanding food deserts, swamps, and mirages as outcomes of food apartheid allows for a reframing of hunger not as scarcity, but as exclusion. The solution is not just charity or supermarket placement, but systemic redress. In Baltimore and other cities, community-led responses from Black urban farming collectives like The Greener Garden and Black Yield Institute are reclaiming agency over land, diet, and health. These groups are not only growing food but reweaving kinship, healing, and resistance into spaces that were deliberately stripped of them. Geography may shape hunger, but community organizing is reshaping the map.

Inflammatory Environments

Urban environments do not merely shape the external conditions of life; they inscribe themselves biologically into the body. The chronic stressors embedded in daily urban living, exposure to pollution, noise, violence, economic insecurity, and poor diet can trigger systemic inflammation, a persistent state of immune arousal that quietly fuels chronic disease. In communities of color, these conditions are neither incidental nor accidental. They are the cumulative outcome of policy decisions, racialized neglect, and built environments designed without health equity in mind. Inflammation, then, becomes more than a biological response; it is a political consequence.

At the cellular level, inflammation is a complex immune response involving cytokines, white blood cells, and hormonal regulation. While short-term inflammation helps the body heal, chronic inflammation (characterized by low grade, persistent immune activation) can damage tissues and increase vulnerability to diseases. Cytokines such as *IL-6* and *TNF-α* play a major role in this process, along with the dysregulation of cortisol, a stress hormone crucial to immune system balance.[11] Studies show that people who live in high-poverty ZIP codes with elevated pollution levels are more likely to experience these inflammatory profiles.[12]

Air pollution's link to asthma is well documented, especially among Black and Latino youth. The GALA II and SAGE II studies, involving over 1400 Latino and 500 Black/African American children, found early-life exposure to nitrogen dioxide (NO$_2$), increased asthma risk by 17% per 5 ppm (95% CL: 1.04-1.31).[13] Similarly, the National Cooperative Inner-City Asthma Study (NCICAS) showed urban pollution exposure (PM$_{2.5}$, NO$_2$, SO$_2$, O$_3$) was significantly associated with increased respiratory symptoms and decreased lung function days later.[14] In neighborhoods such as Mott Haven in the South Bronx dubbed Asthma Alley, Black and Brown children experience ER visits five times about the national average, and 21 times higher than affluent NYC neighborhoods.[15] These disparities reflect environmental racism: minority communities are exposed to 56% more air pollution than their contribution, while white communities breath 17% less than they emit.[16] Research published in the American Journal of Respiratory and Critical Care Medicine further shows that Black residents living in formerly redlined neighborhoods face 3-4x higher risk of uncontrolled or severe asthma than whites in higher-grade areas.[17]

Diet is another key contributor. Highly processed, high-sugar, and high fat diets common in urban food deserts and food swamps promote insulin resistance and stimulate pro-inflammatory cytokine production. Slopen et al. (2016) found that individuals living in food-insecure neighborhoods exhibited elevated markers of C-reactive protein (CRP), a key biomarker of systemic inflammation.[18] For Black and Brown men, the intersection of limited food access, economic precarity, and social stigma around care-seeking intensifies these risks. Communities saturated with fast food marketing and deprived of grocery infrastructure become biologically primed for inflammatory diseases such as diabetes, hypertension, and cardiovascular disease.

Chronic psychosocial stress compounds these physiological burdens. Persistent exposure to neighborhood violence, housing insecurity, systemic racism, and unemployment activates the hypothalamic pituitary adrenal (HPA) axis, sustaining high levels of cortisol, which in turn suppresses immune function over time. These physiological effects are unevenly distributed. A study by Geronimus et al. (2006) on "weathering" shows that Black individuals, especially men show signs of accelerated biological aging due to sustained stress exposure.[19] Inflammatory disease thus emerges earlier and progresses more rapidly in communities subjected to chronic adversity.

Masculinity further shapes how these stressors are embodied. In many Black and Brown communities, masculine identity is shaped by survival imperatives: the need to appear strong, remain emotionally contained, and suppressing vulnerability. While these norms may offer short-term protection in hostile environments, they also encourage behaviors that heighten inflammation and chronic disease. The pressure to "push through" rather than seek care, to self-medicate rather than express distress, has physiological consequences. Over time, this creates conditions ripe for inflammatory diseases, including hypertension, metabolic syndrome, diabetes, and autoimmune dysfunction.[20]

As mentioned earlier in this book, the concept of John Henryism captures this dynamic. Defined as a high effort coping style in response to chronic psychosocial stress, John Henryism is disproportionately observed among Black men navigating racism and economic hardship. Jackson et al. (2010) found that black men who embodied this ethos of "grinding through adversity" were more likely to suffer from high blood pressure, inflammation-related cardiovascular conditions, and depression.[21] Importantly, this pattern was associated with societal expectations of self-sacrifice, strength and silence. The hallmarks of toxic masculinity. Similarly, Berger and Sarnyai (2015) demonstrated that men under chronic psychosocial strain showed marked increases in pro-inflammatory cytokines such as IL-6 and TNF-α, reinforcing the connection between stress, masculinity, and immune dysfunction.[22]

These norms around masculinity also affect how men engage or fail to engage with the healthcare system. Courtenay (2000) argues that traditional masculine behaviors are associated with increased dietary risk, delayed healthcare utilization and avoidance of preventative services.[23] In environments where health-promoting resources are already scarce, like food deserts, neighborhoods with low healthcare density, or areas with high crime, these gendered behaviors intensify vulnerability. Springer and Mouzon (2011) found that Black men who endorse hegemonic masculine norms were significantly more likely to consume fast food, avoid mental health services, and minimize symptoms of physical illness.[24] Within this framework, food and behavior function as coping mechanisms in environments where therapy is stigmatized and self-care is feminized.

The structural implications are staggering. According to the CDC (2022), Black men have the lowest life expectancy of any demographic group in the United States, much of which is attributable to inflammatory diseases such as diabetes, cardiovascular disease, chronic kidney disease, and cancer.[25] Inflammatory coping is not merely a reflection of poor choices' it is the biological echo of living in a social world where masculinity is defined by control, silence, and endurance. In marginalized neighborhoods, where trauma, environmental toxins and food insecurity converge, the masculine ideal becomes entangled with survivalist behaviors, eating to soothe, fighting to protect, and repressing to endure. Inflammation becomes the language through which inequality is written into the body.

Biopolitics of Immunity

The immune system does not operate in a vacuum; it is entangled with the conditions of one's existence. Whoever gets to live in neighborhoods with clean air, anti-inflammatory diets, green space, and access to preventive care is not random, it is structured by power. Michel Foucault's concept of biopower describes how governments and institutions manage populations by regulating the body, deciding who gets to be healthy, who is protected, and who is left vulnerable.[26] For Black and Brown men in urban environments, biopower is not an abstract theory; it is a material reality. It is evident where highways are placed, how police are deployed, where clinics are located, and what food is available. Their bodies become sites of political neglect, shaped not only by exposure to harm but by denial of resources needed to heal. This makes their immune systems political battlegrounds, forced to absorb inequality at the cellular level.

Healthcare access itself reflects these biopolitical dynamics. Preventive services—routine screenings, nutritional counseling, stress reduction, and anti-inflammatory interventions are unevenly distributed across urban spaces. Studies consistently show that predominantly Black neighborhoods have fewer primary care providers, longer wait times, and lower quality facilities than wealthier, white areas. For men navigating rigid masculinity norms that discourage help-seeking, these structural barriers are magnified. The result is delayed diagnosis, advanced diseases at presentation, and higher mortality from conditions that are both preventable and manageable. Immune breakdown is treated as expected rather than preventable, normalized rather than addressed.

These dynamics were laid bare with brutal clarity during the COVID-19 pandemic. Black and Latino men in urban centers experienced disproportionately higher infection, hospitalization, and mortality rates not because of genetic susceptibility, but because of occupational exposure, crowded housing, limited healthcare access, and preexisting inflammatory burden.[27] Immunity during the pandemic functioned as a form of social privilege. The ability to isolate, work remotely, access timely care, and recover safely was unevenly distributed, revealing that immune protection operates as a political resource rather than a universal right.

In practice, biopower plays out through zoning laws, public housing design, infrastructure planning, and criminal legal policy tools that have historically marginalized Black and Brown communities. For example, in cities like Chicago and Detroit, environmental zoning policies have routinely permitted factories and waste processing plans to operate near low-income communities of color, exposing residents to higher levels of air pollution.[28] This leads not only to elevated rates of asthma and respiratory illness, but to chronic immune dysregulation, what public health researchers now recognize as the biological expression of structural violence.[29]

Biopower also functions through what Foucault called disciplinary mechanisms, norms, and institutions that train individuals to regulate themselves. For Black and Brown men, masculinity becomes a mechanism of this control. In environments shaped by surveillance, poverty and incarceration, masculine identity is policed both externally (by society) and internally (through self-discipline). The mandate to appear strong, unemotional, and self-reliant becomes a health hazard. It deters cares-seeking, normalizes suffering, and encourages inflammatory coping behaviors such as alcohol use, overeating, and emotional suppression, all of which weaken immune function.[30]

Importantly, biopower does not only operate through neglect, but it is also enforced through containment and coercion. Incarceration exemplifies this logic. Black and Latino men are disproportionately incarcerated, and prisons are inflammatory environments by design; overcrowded, violent, unsanitary, and trauma-inducing. Upon release, the biological imprint of incarceration, elevated cortisol, suppressed immunity, disrupted sleep is often untreated and ignored further entrenching health disparities.[31] The state's role in managing these bodies, not through care but through control, underscores Foucault's thesis: biopolitics is about who is allowed to live and who is expected to survive without support.

Sperm morphology and function are increasingly recognized as biological mirrors of environmental condition. A 2022 meta-analysis found that exposure to urban air pollution—particularly particulate matter such as PM2.5 and PM10—was significantly associated with decreased sperm count, reduced motility, and a lower proportion of morphologically normal sperm.[32] These effects are not abstract or distant. A population-based study in China observed that men living in highly polluted urban centers exhibited markedly worse sperm quality than their rural counterparts, a disparity linked to heightened oxidative stress and epigenetic alterations in sperm DNA.[33] Importantly, these changes do not simply affect individual fertility; they reflect the biological consequences of prolonged exposure to structurally hazardous environments.

Time-series data further underscore the sensitivity of male reproductive biology to environmental insult. One study demonstrated that short-term spikes in PM2.5 exposure were followed by measurable declines in sperm motility approximately seventy days later—the duration of the spermatogenesis cycle—suggesting a direct temporal relationship between ambient pollution and reproductive impairment.[32] In this sense, sperm becomes a living archive of environmental policy: highways, zoning decisions, industrial siting, and regulatory neglect leave molecular traces within reproductive cells themselves.

Viewed through a biopolitical lens, these findings reveal how environmental governance quietly regulates reproductive capacity. The ability to produce healthy sperm—and by extension, to reproduce without biological penalty—is unevenly protected across urban space. For Black and Brown men disproportionately concentrated in high-pollution neighborhoods, reproductive vulnerability is not the result of personal behavior but of sustained exposure to environments structured by racialized disinvestment. Sperm morphology thus joins immune dysregulation as a marker of political abandonment, demonstrating how inequality is not only lived and survived but biologically transmitted.

Ultimately, the immune system becomes a ledger of political decision making. In Black and Brown men's bodies, inflammation tells a story of centuries of exclusion from clean air, from healthy food, from adequate housing and from dignified care. Their chronic diseases are not just medical; they are the biological residue of state sanctioned neglect. To study immune health in urban Black and Brown men is therefore, to study power, history, and the politics of disposability.

Case Studies of Resistance

Against the backdrop of systemic neglect, racialized food apartheid, and chronic disinvestment, Black and Brown communities across the United States have not only survived—they have innovated. In neighborhoods long treated as biologically expendable, residents have built alternative food systems that resist immune suppression and metabolic harm at their roots. These efforts challenge the very structures that produce health inequity, offering models of care that are collective, culturally grounded, and materially transformative.

Resistance in this context is not only political but biological. It is the reclamation of nourishment, immune protection, and bodily autonomy in environments designed to erode them. Urban farming initiatives, food cooperatives, youth-led programs, and clinic–community partnerships function as counter-biopolitical infrastructures. They interrupt inflammatory exposure, restore access to nutrient-dense foods, and rebuild trust where institutions have failed. Across cities such as Detroit, Los Angeles, Oakland, and Chicago, these initiatives demonstrate that immune justice is achievable when communities control the conditions of care.

Detroit—often cited as a symbol of industrial decline—has emerged as a national epicenter of Black food sovereignty. The Detroit Black Community Food Security Network (DBCFSN), founded in 2006, reframed food justice through a lens of self-determination and land stewardship. Operating D-Town Farm, a seven-acre urban agricultural site on the city's west side, DBCFSN produces organic food, delivers nutrition education, and advances policy advocacy rooted in Black governance. The farm is not simply about food access; it is about repairing generational harm produced by redlining, disinvestment, and nutritional deprivation.[34]

Empirical evidence supports these impacts. Urban agriculture programs have been shown to increase fruit and vegetable consumption, reduce food insecurity, and strengthen social cohesion.[35] In Detroit's North End neighborhood, participation in DBCFSN-affiliated programs reduced food insecurity by approximately 18% over five years.[36] Importantly, the organization explicitly resists racial capitalism by positioning Black residents not as passive recipients of aid, but as producers, decision-makers, and land stewards—reshaping both metabolism and power.

Where corporate grocery chains have withdrawn, cooperatives and mutual aid networks have emerged as vital immune infrastructures. In Chicago, ChiFresh Kitchen—a worker-owned cooperative led by formerly incarcerated Black women—provides culturally relevant meals to schools, shelters, and families in food-insecure neighborhoods. The model integrates employment, restorative economics, and nutrition, addressing both material need and structural exclusion (ChiFresh Kitchen, 2022).[37] By centering dignified labor and collective ownership, ChiFresh disrupts the link between poverty, stress, and inflammatory disease.

During the COVID-19 pandemic, mutual aid networks in cities such as New York and Atlanta further illustrated the life-saving role of community governance. These groups pooled resources, bypassed institutional gatekeepers, and distributed fresh food to thousands of households weekly, often reaching families excluded from formal aid systems.[38] In communities where trust in state and healthcare institutions was deeply eroded, these networks functioned as immune lifelines. As Ruth Wilson Gilmore argues, such efforts create "abolitionist geographies"—spaces where care replaces abandonment and survival is reorganized around interdependence rather than profit.[39]

Youth-led initiatives have also become central to food-based resistance, linking nourishment to political education and long-term health. In Oakland, Acta Non Verba (ANV) operates a youth urban farm where children and adolescents grow food, manage a community-supported agriculture (CSA) program, and reinvest profits into education. The program integrates climate justice, nutrition science, and cultural affirmation, cultivating both metabolic health and political consciousness.[40]

Similarly, Grow Greater Englewood in Chicago transforms abandoned lots into gardens and green corridors led by Black youth. Participants engage in food distribution, neighborhood advocacy, and entrepreneurship. Evaluations of the program report improved dietary behaviors, reductions in body mass index (BMI), and increased self-esteem among youth participants.[41] These initiatives demonstrate that food justice is not merely about consumption, but about agency—teaching young people to see their bodies, neighborhoods, and futures as worthy of care.

Healthcare institutions have also begun, albeit unevenly, to partner with community organizations in addressing diet-related inflammatory disease. In Baltimore, the FoodRx initiative connects primary care clinics with local farms and markets, providing weekly produce prescriptions to patients with hypertension and diabetes. Early evaluations show meaningful reductions in blood pressure and improved glycemic control over six months.[42] Similar programs in Boston and Minneapolis have reported 10–12% decreases in household food insecurity and improved adherence to clinical care.[43] These interventions illustrate how medicine, when decentered from extractive models, can function as a tool of repair rather than control.

Perhaps no figure better captures the symbolic and practical dimensions of food resistance than Ron Finley of South-Central Los Angeles. By transforming the strip of land outside his home into a garden (an act initially deemed illegal), Finley catalyzed a movement. The Ron Finley Project now trains residents to grow food, reclaims abandoned lots, and reframes gardening as political empowerment. "Growing your own food is like printing your own money," Finley argues, underscoring food sovereignty as economic and biological autonomy.[44] Evaluations of the project show that participants increased fruit and vegetable consumption by approximately 40% and reported improved family meal preparation habits. Yet the deeper impact lies in the shift from dependency to agency—from biological vulnerability to collective strength. As Finley notes, "I'm planting the seeds of freedom."

Taken together, these case studies illustrate a crucial counterpoint to immune injustice: when communities reclaim land, food, and care, they interrupt the inflammatory pathways produced by structural neglect. These initiatives do more than improve nutrition; they rebuild trust, reduce chronic stress, and restore the conditions under which immune systems can function optimally. In doing so, they expose a central truth of urban health: immunity is not simply a biological outcome, but a political achievement.

Conclusion: Immune Justice as Urban Justice

Inflammation is not merely a biological response; it is a biological register of inequality. Throughout this chapter, we have traced how urban infrastructures, policy decisions, and structural racism create environments where immune systems are pushed into overdrive. From polluted air and heat islands to ultra-processed food environments, chronic surveillance and medical neglect, the urban landscape becomes a site where bodies, especially Black and Brown men's bodies, bear the cost of systemic abandonment. Immunity, in this perspective, is neither neutral nor natural. It is a political outcome shaped by where one lives, works, and eats and is allowed to rest.

Masculinity, too, is entangled in this physiology of inequity. For many Black and Brown men, the expectation to be resilient, unemotional, and self-sufficient (developed as survival strategies under racial capitalism) not only delays care but compounds biological stress. The very traits celebrated as strength in communities navigating systemic hardship can become liabilities when they prevent early intervention, community healing, or vulnerability. The body interprets these social scripts of stoicism, of silence, or survival and translates them into suppressed immunity, heightened inflammation, and increased risk for chronic disease. What we call masculinity, the body registers as cortisol, cytokines, and blood pressure.

To confront inflammation is to confront injustice. Immune health cannot be reduced to individual behavior or personal responsibility when exposure itself is unequally distributed. Addressing immune inequity requires policy interventions that expand green infrastructure, regulate urban zoning to prevent food swamps, remediate environmental toxins, and subsidize access to nutrient-dense foods in historically disinvested neighborhoods. It requires redefining health as a collective outcome shaped by place, power and political will not merely by choice. It means recognizing that immune health is a matter of housing justice, environmental justice, and racial justice. Furtherly, it means building health systems that affirm, not pathologize, the emotional and cultural realities of Black and Brown men's lives.

Crucially, this chapter has also demonstrated that immune vulnerability is not destiny. As the case studies in the previous section illustrate, communities have long resisted biological abandonment through collective care. Urban farms, food cooperatives, youth-led nutrition movements, and clinic-community partnerships represent more than service delivery models; they are counter biopolitical infrastructures. By reclaiming land, nourishment, and governance, these initiates reduce inflammatory burden, restore trust, and reassert the right to health as a shared good rather than a market privilege.

These grassroots responses remind us that immunity is not only about defense, it is account connection, regeneration, and survival. They challenge dominant health paradigms that individualize risk while ignoring structural exposure. In doing so, they reveal a fundamental truth: immune systems do not fail in isolation, they falter in environments shaped by neglect, extraction, and inequity. When those environments are transformed, immune possibility expands.

Immune justice, then, is urban justice. It is the right to breathe clean air, to eat nourishing food, to experience care without stigma, and to inhabit a body not constantly pushed into crisis. And like all struggles for justice, it does not emerge from policy alone, but from the margins—from gardens grown into concrete, from kitchens turned into cooperatives, and from communities refusing to accept premature illness as inevitable. Justice, as always, grows in the cracks.

Chapter 7: Concrete Wombs and Metal Cradles

Concrete Wombs and Metal Cradles—Fatherhood, Care, and Inherited Wounds

> "Some carry children in their arms. Others carry them in their scars."- Anonymous.

To be a father in the concrete corridors of America's cities is to carry life and rupture, to nurture in spaces never meant for softness. For Black and Brown men, fatherhood often begins not in still hospital rooms but in neighborhoods cracked by policy and patrolled by fear. The metaphor of concrete wombs and metal cradles captures the tension between creation and confinement, between the desire to protect and the limits imposed by systemic neglect. Concrete wombs: the neighborhoods that birth children into systems of disinvestment, redlining, and surveillance. Metal cradles: the carceral state, juvenile facilities, and school to prison pipelines that too often rock Black and Brown boys to sleep with institutional hands. These environments do not merely shape parenting practices; they shape stress physiology, immune regulation, and the biological conditions under which care is transmitted. This is the geography of fatherhood for many men, where love must be taught alongside survival, and where care is a radical act of resistance.

In these environments, masculinity is shaped as much by scarcity as by strength. Fatherhood becomes both a burden and a blessing, a site of renewal and a source of risk. Social narratives have long reduced Black and Brown fathers to tropes, absent, dangerous, disposable. Yet behind these distortions are men walking children to school while dodging court summons, working night shifts to provide without being present, or learning love with little blueprint for vulnerability. These are the fathers born of concrete wombs, men raised in fractured systems, still trying to build something whole. Their cradles may be metal, marked by police contact, economic precarity, or health disparities but their commitment to care survives, often in quiet unrecognized ways. This chapter examines how Black and Brown men practice fatherhood under pressure: how they pass down not just trauma but tenderness; not just wounds, but wisdom.

The Labor of Fatherhood
"I held my son in my arms for the first time in a hospital room that smelled like bleach and quiet sorrow. Outside, helicopters chopped the sky and police sirens rushed through neighboring streets, and inside, I promised him peace I wasn't sure the world could give."

That's how Grayling, a 33-year-old father from Charleston, SC described the birth of his child. He wasn't only recalling a memory; unbeknownst to him he was offering a parable about love in hostile terrain, about the weight of care in a world that often denies men like him the right to be tender. For many Black and Brown men, fatherhood begins in places marked not by safety but by surveillance. The labor of fatherhood is not merely emotional or economic, it is existential. It is the work of raising a child while navigating systems that are indifferent or actively hostile to your survival. It is love expressed in layers of armor.

Fatherhood, particularly in marginalized communities, is more than a biological role or social identity; it is a form of labor that unfolds under constraint. This labor is physical, working long hours in precarious jobs to provide material stability. It is cognitive, anticipating danger, negotiating institutions, and making daily calculations about safety, discipline, and exposure. It is emotional, absorbing fear, frustration, and exhaustion so that children can experience moments of calm. And it is political. To choose to stay, to protect, to nurture, and to model care in a society structured to criminalize and discard Black and Brown men is an act of quiet defiance.

Fatherhood in marginalized communities is more than a biological role. It is a political act. To choose presence, protection, and nurturance in a social order that systematically undermines Black and Brown men's health, mobility, and longevity is a form of defiance. Contrary to dominant narratives that frame these men as absent or disengaged, research consistently demonstrates that Black fathers exhibit diverse and meaningful patterns of involvement, including high levels of emotional engagement, caregiving, and responsibility, even under conditions of structural constraint.[1] These men often father within institutions that make caregiving extraordinarily difficult: unstable labor markets, punitive child support systems, untreated trauma, and mass incarceration, which has been shown to fracture father identity and disrupt paternal involvement long after release.[2] And yet, they do father: in the streets, in living rooms, through school drop-offs, late-night phone calls, on FaceTime from prison phones, and in memories passed down like heirlooms. Fatherhood persists even when presence is policed and care is constrained.

At the core of this labor lies a sharp and enduring tension: the tension between protection and vulnerability. How does a man shield his child when he himself is unprotected? What does it mean to be the guardian while still carrying the wounds from one's own upbringing? These are not abstract questions; they are embodied daily by men guiding children through public schools, courtrooms, clinics, and neighborhoods shaped by violence and neglect. In environments where resources are scarce and trauma is ambient, fatherhood becomes a continuous balancing act between strength and softness, provision and presence, silence, and emotional truth.

Masculinity complicates this labor. Social expectations that men remain stoic, self-reliant, and emotionally contained often shape how fatherhood is performed and perceived. While these traits may function as protective armor in hostile environments, they can also become barriers to help-seeking, vulnerability, and relational care. Contemporary scholarship shows that hegemonic masculine norms often conflict with nurturing fatherhood, discouraging emotional expression and reinforcing internalized stress.[3] At the same time, emerging research on "caring masculinities" demonstrates that engaged fathering can reshape masculine identity itself, expanding men's emotional repertoires and fostering relational strength rather than domination.[4] This suppression is itself work, labor that takes a physiological toll. The body registers these demands through heightened stress, disrupted sleep, and cumulative wear, translating social responsibility into biological strain. In this way, fatherhood becomes an embodied practice: care is metabolized, vigilance becomes hormonal, and love leaves traces in the body.

This chapter does not seek to redeem or romanticize fatherhood. It seeks to understand it in context. For Black and Brown men living in cities marked by surveillance, scarcity, and structural abandonment, fatherhood is both a tender rebellion and a political act. In places where institutions fail to nurture, these men build cradles from scrap, protect futures with calloused hands, and pass on not just pain, but survival, pride, and memory. "Concrete wombs" hold generations born into confinement, systems that regulate life before it even begins. "Metal cradles" evoke the carceral institutions, detention centers, and disciplinary systems that rock Black and Brown boys into adulthood with institutional hands. Yet within these conditions, care persists.

What follows is not just a study of fatherhood, it is an excavation of legacy, of inheritance both biological and emotional, of labor both visible and unseen, and of how care endures under pressure. In the chapters ahead, fatherhood will emerge not as a footnote to masculinity, but as one of its most radical reimaginings.

Fatherhood in the Margins: Masculinity Reimagined

In the national imagination, the phrase *"Black fatherhood"* is too often synonymous with absence—a man missing from the household, emotionally detached, or physically gone. This racialized stereotype has functioned as one of the most enduring myths in American discourse, repeatedly invoked to explain poverty, crime, and social disorder. Yet this narrative persists not because it reflects reality, but because it absolves systems (e.g., racism, mass incarceration, labor exploitation) of responsibility. It privatizes structural failure, blaming men for being "missing" from homes they were systematically excluded from.

Contemporary data continues to disrupt this mythology. Analyses drawing on the CDC's National Survey of Family Growth demonstrate that Black fathers are as involved, and in some domains more involved, in daily caregiving activities—such as bathing, feeding, and playing with children—than white fathers, regardless of residential status.[5] These findings challenge the assumption that co-residence equals care and reveal how dominant measures of fatherhood systematically undervalue relational labor, especially when performed by marginalized men.

Qualitative research deepens this corrective. Recent interviews with low-income Black and Latino fathers show that many define fatherhood not primarily through financial provision, but through emotional presence, moral guidance, and relational accountability.[6] Fathers describe attending school events when possible, maintaining daily phone calls across distance, offering counsel, and modeling resilience in the face of constraint. In these accounts, emotional availability emerges as a form of resistance—an insistence on humanity in systems that repeatedly deny it.

Yet the economic burden on marginalized fathers remains heavy. Structural unemployment, wage gaps, and the erosion of stable union jobs have made traditional provider roles difficult, if not impossible, for many men in urban communities. Unemployment among Black men consistently doubles that of white men, and in some urban areas, jobless rates for young Black men reach as high as 50%.[7] For men whose masculine identity has been tethered to provision, this exclusion generates profound "role strain", a psychological conflict between the expectations of manhood and the constraints of structural inequality.[8]

This strain is not merely psychological; it is embodied. When masculinity is defined through control, independence, and economic dominance, structural exclusion becomes an existential crisis. Scholars describe this tension as producing internalized shame, withdrawal, depressive symptoms, or compensatory hypermasculine behaviors.[9] These responses are often misread as personal failure, rather than what they are: adaptations to systemic betrayal.

Incarceration intensifies these dynamics. The U.S. carceral system not only physically removes fathers from their families, but also socially and economically erases them upon return—blocking access to employment, housing, and public assistance. Black men are incarcerated at nearly six times the rate of white men, while Latino men are nearly three times more likely to be imprisoned.[10] Nearly one in nine Black children has an incarcerated parent, most often a father, compared to one in 28 white children.[11] Incarceration disrupts not just physical presence but economic stability, emotional bonding, and the intergenerational transmission of care. Families left behind absorb the trauma of forced absence, often under condition of economic precarity and social stigma. For many Black and Brown men, incarceration is not an exception but a rite of passage, a structural "norm" that disrupts the possibility of consistent fatherhood. Yet even behind bars, fatherhood persists. Programs like the Fathers and Children Together (FACT) initiative have shown that men incarcerated in state prisons often maintain bonds through letters, visits, storytelling, and intentional emotional work.[12] These practices—largely invisible to policymakers function as lifelines, preserving dignity and continuity where the state offers none.

Across these contexts, a different portrait of masculinity emerges. One defined not by domination, but by dedication. Not by stoicism alone, but by care under pressure. Fatherhood becomes a site where masculinity is not abandoned but reworked, expanded to include emotional labor, vulnerability, and relational presence. This aligns with contemporary scholarship on *caring masculinities*, which emphasizes interdependence, nurturing, and accountability over control.[13]

Crucially, this reimagining occurs in the margins. It unfolds in grocery store aisles, visitation rooms, public housing stairwells, and late-night conversations where children ask questions men were never taught how to answer. These are not abstract transformations; they are daily practices forged under constraint. Masculinity, in this sense, is not a fixed identity but a living process shaped by context, responsibility, and love. Fatherhood in the margins reveals masculinity as potential rather than pathology. It shows how Black and Brown men, even when structurally sidelined, continue to perform care in ways that disrupt dominant hierarchies of value. In reimagining fatherhood, they also reclaim masculinity, not as economic dominance or emotional withdrawal, but as presence, intention, and the quiet insistence on connection in hostile environments.

Care as Resistance

Fatherhood in Black and Brown communities, particularly within the urban United States, often unfolds under the weight of structural oppression and cultural misrepresentation. Yet, amid narratives that pathologize marginalized fatherhood, caregiving emerges as a form of political and cultural resistance. Through daily acts of presence, guidance, and intergenerational responsibility, men resist the racialized tropes of absenteeism and redefine masculinity through care. In contexts where survival has long been prioritized over softness, care itself becomes defiant. This section examines how fatherhood, mentorship, and communal spaces like barbershops and fraternal organizations serve as transformative arenas of resistance, wellness, and social belonging.

Black barbershops, long-standing institutions within Black communities, exemplify this caregiving resistance. Historically, places of cultural exchange and political dialogue, barbershops have increasingly been recognized as critical venues for health promotion and emotional support. A landmark study by Releford et al. (2010) demonstrated that barbershops could effectively serve as sites for cardiovascular health interventions, with trained barbers promoting blood pressure screenings among African American men.[14] More recently, a randomized controlled trial by Victor et al. (2018) showed that barbershop-based interventions—when paired with pharmacist-led medication management—produced an average reduction of 21.6 mmHg in systolic blood pressure among Black male clients over six months.[15] These findings underscore that care delivered in trusted, culturally grounded spaces can succeed where traditional medical systems often fail.

Beyond measurable health outcomes, barbershops function as relational sanctuaries. Qualitative research highlights their role as spaces of identity affirmation, mentorship, and emotional expression. Hart and Bowen (2004) found that Black men often experience barbershops as environments of trust and confidentiality, where conversations extend beyond grooming to encompass fatherhood, grief, survival, and hope.[16] In these chairs, masculinity is renegotiated—not through dominance or emotional restraint, but through listening, storytelling, and mutual support. Linnan et al. (2011) similarly describe barbershops as "safe emotional outlets," particularly for men unlikely to engage formal mental health services.[17] Barbers are frequently viewed as trusted lay counselors, capable of recognizing distress and facilitating pathways to care without invoking stigma or emasculation.

Care as resistance also extends beyond informal spaces into organized networks of mentorship and communal responsibility. Black Greek Lettered Organizations (BGLOs), particularly Alpha Phi Alpha Fraternity, Inc., represent another powerful expression of fatherhood beyond biological lineage. Founded on December 4, 1906, Alpha Phi Alpha has long emphasized scholarship, service, and leadership as tools of racial uplift. Through initiatives such as *Go-to-High School, Go-to-College* and *Project Alpha*, the fraternity mentors young men while modeling forms of masculinity grounded in accountability, emotional presence, and collective responsibility.[18] These organizations (Alpha Phi Alpha Fraternity, Inc., Kappa Alpha Psi Fraternity, Inc., Omega Psi Phi Fraternity, Inc., Phi Beta Sigma Fraternity, Inc. and Iota Phi Theta Fraternity, Inc.) function as infrastructures of care, filling institutional voids created by underfunded schools, fragmented families, and systemic disinvestment.

Taken together, these practices reveal a reimagining of masculinity not as control or emotional suppression, but as stewardship and connection. In a society that frequently frames Black and Brown men as threats, absentees, or liabilities, these men cultivate lives of presence—raising children, mentoring youth, and sustaining communal bonds. Their labor often remains invisible in economic metrics and policy discourse, yet it is deeply felt in barbershops, fraternity houses, classrooms, and neighborhood streets. Here, care is not incidental; it is intentional, relational, and political. In choosing to nurture where neglect is expected, Black and Brown fathers enact resistance not through protest alone, but through the radical, everyday work of love.

Biological Imprints: The Epigenetics of Fatherhood

Fatherhood leaves a biological trace—not just in DNA, but in the way, genes are expressed across generations. While we often think of inheritance as fixed, emerging research in epigenetics reveals a more dynamic reality: experiences such as stress, trauma, and care can influence biological signaling pathways that affect offspring development. Epigenetics refers to modifications in gene expression that occur without changing the underlying DNA sequence, often through mechanisms such as DNA methylation, histone modification, and non-coding RNA activity. In this way, the emotional and environmental conditions shaping a father's life, particularly those marked by racism, economic precarity, violence, or chronic stress can become biologically legible in the bodies of his children.

Paternal transmission of stress-related biological signals is an expanding area of inquiry with profound implications for understanding intergenerational health. Early experimental work demonstrated that stress exposure in male animals prior to conception altered offspring stress reactivity and neurodevelopment through epigenetic changes in sperm.[19] Human studies have since begun to echo these findings. Fathers who experienced early life adversity including neglect, abuse, or sustained socioeconomic stress have been shown to exhibit distinct sperm DNA methylation profiles associated with genes involved in stress regulation and brain development.[20] These epigenetic patterns have been linked to altered cortisol responses and emotional regulation in offspring, suggesting that paternal experience before conception may shape children's vulnerability to stress-related disorders.

Cortisol plays a central role in this process. Chronic exposure to stress, whether from systemic racism, housing instability, incarceration, or unemployment often leads to prolonged dysregulation of the hypothalamic–pituitary–adrenal (HPA) axis in men. Elevated or blunted cortisol rhythms not only compromise immune, cardiovascular, and metabolic health in fathers themselves, but also appear to influence epigenetic signaling in sperm. Research on trauma-exposed populations has shown that children of men who experienced severe or chronic stress may display altered cortisol profiles, increasing susceptibility to anxiety, depression, and inflammatory disease later in life (Yehuda and Bierer 2009).[21] In this regard, stress does not end with the individual—it reverberates biologically across generations.

Recent large-scale human studies have significantly refined this understanding. In a longitudinal cohort analysis, Kretschmer et al. (2023) found that fathers exposed to chronic psychosocial stress prior to conception exhibited altered DNA methylation patterns in sperm linked to neurodevelopmental and immune-regulatory genes. These epigenetic signatures were associated with heightened stress reactivity and emotional vulnerability in offspring during early childhood, even after accounting for maternal stress and socioeconomic factors.[22] Crucially, the authors emphasize that these effects represent *probabilistic risk*, not biological determinism, highlighting sensitivity rather than inevitability.

Environmental exposure further complicates these pathways. Wu et al. (2022) demonstrated that paternal exposure to urban air pollution was associated with oxidative stress, disrupted sperm chromatin integrity, and altered microRNA expression—mechanisms increasingly recognized as critical to early embryonic development.[23] These findings suggest that environmental conditions common in disinvested urban neighborhoods—pollution, heat, and toxic exposure—may shape sperm-mediated signaling involved in immune programming and stress regulation, linking environmental injustice directly to reproductive health.

Most recently, Margiana et al. (2025) synthesized emerging evidence across epigenetics, developmental psychology, and public health to propose an integrative model of paternal transmission.[24] Their framework emphasizes that intergenerational risk operates through multiple, interacting channels: sperm epigenetic modification, altered seminal fluid composition, prenatal stress buffering, and postnatal caregiving environments. For Black and Brown fathers, whose lives are disproportionately shaped by structural stressors, this model underscores how inequality becomes embodied—but also how intervention remains possible at multiple points. Biology, in this view, is not fate but responsiveness.

Importantly, fathers transmit more than stress. Nurturing involvement also leaves biological imprints, often protective ones. Children whose fathers are emotionally present and consistently engaged demonstrate improved stress regulation, lower rates of depression, and stronger immune function. Father involvement, defined not simply by physical presence but by emotional attunement and co-regulation, has been associated with lower salivary cortisol levels and reduced externalizing behaviors in children.[25] Longitudinal studies further indicate that consistent paternal support is linked to lower levels of inflammatory markers such as C-reactive protein and interleukin-6, both predictors of chronic disease risk.[26]

Father involvement also plays a critical role in shaping mental health outcomes. In a large longitudinal study, Sarkadi et al. (2008) found that children with engaged fathers were less likely to develop emotional and behavioral problems, even after controlling for maternal involvement and socioeconomic status.[27] These protective effects were especially pronounced in high-risk populations, including low-income families and communities of color. When fathers were involved in day-to-day care and communication, children exhibited greater resilience to adversity and were less likely to report anxiety, depressive symptoms, or poor academic performance. Qualitative research adds texture to these findings. In interviews with Black and Latino fathers in low-income neighborhoods, many described their roles not just as providers, but as emotional anchors within the family. Despite barriers like incarceration or underemployment, many fathers emphasized the importance of "being there," emotionally if not always physically. For them, caregiving was a form of resistance, a deliberate effort to break intergenerational cycles of silence, absence, and pain through emotional presence, mentorship, and relational labor, even under conditions of incarceration, economic exclusion, and social surveillance.[28]

Taken together, this body of research reframes fatherhood as both a social and biological role. What fathers endure, suppress, or heal can shape their children's emotional and physical development at the molecular level. Yet this knowledge does not indict fathers; it indicts environments. For men navigating chronic adversity, the question is not whether biology is being shaped, but whether it will be shaped by isolation or connection, stress, or care. Acts of love—gentle touch, listening, consistency, emotional presence are not merely symbolic. They are biological interventions, capable of softening inherited stress and rewriting the molecular scripts passed from one generation to the next.

Sidebar: What We Know—and Don't Yet Know—About Paternal Epigenetic Inheritance
While evidence for paternal epigenetic transmission is growing, important scientific limitations remain. Much of the strongest causal evidence comes from animal models, where environmental exposures can be tightly controlled across generations.[29] Human studies, by contrast, often rely on observational designs, making it difficult to fully disentangle biological transmission from shared social environments,

parenting practices, or postnatal stress exposure.[30] These methodological constraints complicate claims of direct causality and require careful interpretation.

Additionally, epigenetic marks in sperm such as DNA methylation and small non-coding RNAs may be dynamic rather than permanent. Evidence suggests that many epigenetic signals are sensitive to timing, context, and subsequent environmental exposures, raising questions about their stability across the life course and across generations.[31] As a result, paternal epigenetic inheritance is best understood not as a deterministic mechanism, but as one component of a broader biopsychosocial transmission process.

These limitations do not negate the relevance of paternal epigenetics; rather, they clarify its scope. Current evidence supports the conclusion that paternal stress, trauma, and care shape offspring biology through *interacting biological and social pathways*, not isolated genetic determinism.[32] In this sense, epigenetics does not replace structural explanations—it reinforces them. The emerging science underscores how bodies respond to inequality, while leaving ample room for social intervention, healing, and change.

Carceral Fatherhood

I write this section not as someone who has known incarceration firsthand, but as someone who has seen how its shadows stretch across lives, especially in Black and Brown communities. The data are sobering, but even more haunting are the silences: the letters unsent, the visits missed, the birthdays remembered only through bars and bulletproof glass. These absences are not merely logistical. They are cellular, generational, and deeply embodied in the lives of the men and children they separate.

As of 2024, nearly five million U.S. children have experienced the incarceration of a parent at some point in their lives, with one in seven Black children currently living with a parent behind bars, compared to one in twenty-eight Latino children and one in sixty white children.[33] Black men continue to be incarcerated at a rate nearly six times higher than white men.[34] In some states, more than 40 percent of Black men aged 18–35 have had contact with the criminal legal system.[35] These figures reflect far more than individual behavior; they are the cumulative result of systemic inequities—racially biased policing, mandatory minimum sentencing, cash bail regimes, and decades of disinvestment in urban communities.

For fathers behind bars, caregiving becomes fractured, mediated by surveillance, institutional rules, and profound emotional strain. Many attempt to remain present through letters, monitored calls, and virtual visits, such acts of care compressed into narrow windows of access. Jamarr, 22 shared a letter his father incarcerated in Atlanta wrote, "Every night I replay our last visit, how my son wouldn't let go of my neck when the time came to leave. I'm still holding that hug." His words echo across a generation of men learning how to love from confinement. These efforts are often invisible to the public eye, yet they are essential lifelines for both father and child.

The emotional tax of incarceration weighs heavily on father child relationships particularly between fathers and sons. When a father is removed from the home and no stable male caregiver— an uncle, grandfather, mentor is present, sons often turn to the external neighborhood as a surrogate structure. These "metal cradles," forged from peers, block associations, and street codes, frequently lack the emotional literacy or protective guidance of stable caregiving. As Elijah Anderson (1999) described in *Code of the Street*, urban boys often come of age navigating "street" versus "decent" codes without the grounding presence of a father to model resilience, care, and regulated masculinity.[36] The result is often a gendered survivalism that valorizes stoicism and suppresses grief.

Quantitatively, the fallout is measurable. Children of incarcerated parents are six times more likely to experience behavioral problems, and twice as likely to suffer from depression and anxiety compared to peers without incarcerated parents (Turney, 2022).[37] One in three Black boys with incarcerated fathers demonstrates early-onset aggression, difficulty in school, and emotional withdrawal.[38] For incarcerated men themselves, the environment worsens pre-existing health disparities. Studies document elevated rates of hypertension (38%), diabetes (11%), and depression (over 40%), conditions often exacerbated by overcrowding, poor air quality, inadequate healthcare, and chronic stress.[39]

Reentry into society, while framed as a return to freedom is often its own trauma infused chapter of fatherhood. Formerly incarcerated men must reestablish trust, reconnect emotionally, and navigate economic hardship, stigma, and systems that often block employment and housing. These barriers are not just social, they are biological. Recidivism is not only a legal or economic outcome; it is often preceded and followed by deteriorating health. Chronic stress, untreated mental illness, and lack of access to healthcare contribute to a revolving door of incarceration that disproportionately impacts Black and Brown men. A study by Wang et al. (2020) found that formerly incarcerated individuals with unmanaged chronic conditions such as hypertension and substance use disorders were significantly more likely to be reincarcerated within three years.[40] For Black and Latino men, the intersection of medical neglect, trauma, and structural racism results in elevated cortisol levels, disrupted sleep, and inflammatory profiles that mirror those of combat veterans.[41]

Statistically, over 60% of individuals released from prison have at least one chronic health condition, and among Black men, rates of diabetes, asthma, and depression surge after releasing due to lack of continuity of care.[42] This biological vulnerability compounded by food insecurity, housing instability, and limited access to health insurance becomes a key driver of recidivism. Rather than a lack of will, it is often a failure of the system to treat health as foundational to reentry success. These men return to environments that trigger the very stress responses that incarceration exacerbated, making it difficult to stay healthy, stay employed, and stay free.

Yet amid this harm, many fathers describe reentry as an opportunity to *re-father*—to offer new scripts of masculinity rooted in emotional presence, patience, and accountability. A 2021 qualitative study by Mitchell and Davis found that Black fathers often described post-release fatherhood as "walking back into a world that didn't wait."[43] Their children had learned to survive without them, and rebuilding bonds felt both urgent and fragile. One young man, reflecting on his father's incarceration, shared: "When my dad went away, I was nine. I acted like it didn't matter, but it did. I started copying how the older boys on the corner moved. That's who taught me how to be a man—until I realized I didn't want to end up like them."[43] His words capture the vacuum incarceration creates and the limits of the street as a substitute for paternal care.

Fatherhood from behind walls is still fatherhood; constrained, reshaped, but often deeply intentional. It resists erasure and demands recognition. It compels us to ask a harder question: what would our neighborhoods look like if systems were designed to support, rather than separate, Black, and Brown families? What if the cradle wasn't metal, and the womb wasn't concrete?

Note: Incarceration statistics are updated periodically. Figures cited reflect the most recent national data available at time of writing (2023-2025).

Fathers of the Block, Not Just the Bloodline

Fatherhood in marginalized communities does not always flow through bloodlines. In neighborhoods marked by structural absences—of fathers, of opportunity, of sustained state investment—care is often remixed into something collective. This is the labor of community fatherhood: men who take on caregiving roles for youth they did not biologically conceive but whom they nurture, guide, and protect. These men are barbers, coaches, teachers, mentors, and civic fathers who transmit wisdom, offer stability, and model alternative ways of being. In places where incarceration, premature death, or economic displacement have destabilized nuclear families, these men become anchors—holding youth steady in environments designed to unsettle them.

As I write this section, I think of Mr. Leroy Lewis, the long-time director of the College of Charleston's Upward Bound program. I was raised by a single mother, and it was Mr. Lewis who stepped into a role that quietly transcended professional obligation. He taught me how to tie a tie and the difference between a salad fork and a dessert fork—not as performance, but as ritual. These lessons were about dignity, about belonging in spaces that were not built with boys like me in mind. He modeled emotional intelligence long before I had language for it, showing how to navigate hostility not with force, but with restraint, reflection, and composure. He taught me that masculinity could be caring, emotionally literate, and grounded. His impact was not abstract. It was deeply practical and in hindsight, profoundly biological. In moments when rage or silence might have turned inward, his presence recalibrated my nervous system. He offered co-regulation before I knew the term, modeling calm in a world that often-demanded vigilance. That kind of care does not simply shape character; it shapes stress responses, emotional regulation, and long-term health trajectories. It is the quiet labor of preventing harm before it hardens.

Research affirms what stories like mine suggest. Non-biological mentors significantly buffer Black and Brown boys from the psychological, behavioral, and health consequences of adverse environments. A longitudinal study by Raposa et al. (2019) found that adolescents engaged in formal mentoring programs experienced reductions in depressive symptoms, substance use, and behavioral problems, particularly when mentors modeled emotional regulation and future orientation.[44] Among Black youth specifically, DuBois and Silverthorn (2005) reported that consistent, culturally affirming mentorship reduced school dropout rates and improved self-esteem and peer relationships, social determinants with clear downstream health implications.[45]

More recent data from the National Mentoring Resource Center (2020) shows that structured mentorship programs are associated with a 32% improvement in emotional regulation, a 40% reduction in school truancy, and significant decreases in risk-taking behaviors among youth of color.[46] Qualitative studies echo these findings. In an ethnographic study of community mentorship in Chicago, Jarrett et al. (2011) found that older Black men serving as "neighborhood uncles" mitigated the psychological effects of father absence through informal conversations, accountability, and role modeling—creating what the authors described as a form of *social immunization* against environmental stressors.[47]

These informal kinship networks do not only buffer youth; they also function as wellness ecosystems for the men who participate in them. When men engage in caregiving, whether through coaching, mentoring, or informal fathering, their own health trajectories often improve. Griffith et al. (2014) argue that caregiving roles allow African American men to redefine masculinity away from stoicism and self-destruction and toward connection, purpose, and relational accountability. In their study of men in Detroit, participants who served as mentors reported higher life satisfaction, reduced perceived stress, and greater engagement in health-promoting behaviors such as exercise and routine medical care.[48]

Community fatherhood, then, is reciprocal medicine. It challenges dominant narratives that frame Black and Brown men as absent or emotionally disengaged, instead illuminating the vast, often invisible care labor they perform. It is also a direct resistance to carceral and economic logics that define fatherhood solely through financial provision or legal proximity. These men offer what the state has withheld: presence, consistency, and emotional grounding. They teach boys how to breathe through anger, how to reflect before reacting, how to cry without shame.

In doing so, they help reshape what is inherited by the next generation, not only through genes, but through gestures. Through tone. Through calm hands and steady voices. Fathers of the block remind us that biology is not destiny, and neither is bloodline. Care travels where it is needed most, carried by men who understand that to father is not merely to reproduce, but to protect, to guide, and to stay.

Conclusion: Wounded Healers, Radical Care

To father is to labor, to love, and to leave something behind. In the margins of urban America—between systemic neglect and generational pain—Black and Brown men have long performed an unrecognized form of carework. This labor is rarely named, rarely resourced, and often misunderstood. Yet it is here, in conditions not designed for tenderness, that many men practice a radical form of love. Fatherhood, in this context, is not merely biological. It is political. It is a refusal to disappear. Across this chapter, we have seen how fatherhood unfolds under pressure: in households shaped by unemployment and surveillance, in relationships stretched by incarceration and separation, and in communities where care is often improvised rather than supported. Still, men show up. Through intentional fathering, mentorship, and community care, they disrupt the narratives that have long painted Black and Brown fathers as absent or disposable. What emerges instead is a portrait of men carrying wounds while actively working to ensure those wounds are not the only inheritance passed on.

This labor is not symbolic; it is embodied. Research consistently shows that children with engaged fathers (biological or social) experience better mental health, stronger academic outcomes, and fewer behavioral challenges.[49] Yet the opportunity to father fully is unevenly distributed. Structural barriers—unemployment, housing instability, carceral separation, and inadequate social support continue to fracture families and constrain men's ability to care. These ruptures harm children, but they also harm men, depriving them of one of the most powerful pathways to healing: relational connection.

If care is where healing begins, then policy must meet fatherhood with seriousness. Paid family leave, accessible mental health services, trauma-informed reentry programs, and stable housing are not luxuries—they are infrastructure for care. Community-based fatherhood initiatives demonstrate what becomes possible when men are supported rather than surveilled. When fathers are given space to be present, emotionally whole, and economically stable, cycles of harm begin to loosen. Masculinity itself starts to change shape.

This chapter is dedicated to the wounded healers—the men who modeled care when no blueprint existed. Linard McCloud, Dr. Robert Perrineau, Alan Smith, Dr. Maurice Cannon, Talim Lessane, Antonio Robinson, Luther Haynes, Jonathan Bass, Charlie Shedrick, and my biological father, Dexter Nelson. In your presence, I learned that masculinity is not dominance but devotion, not silence, but stewardship. You taught me how to remain intact in a world that demanded fragmentation. Your care was not loud, but it was enduring. Not performative, but precise. It was not just memory, it is legacy. Black and Brown fathers are not anomalies. They are architects of continuity. Their care is often carried quietly, often under duress, and is among the most radical forces shaping the future of urban life. To recognize them is not sentimentality. It is truth.

Chapter 8: From Streets to Studies

From Streets to Studies—Reclaiming Research, Reclaiming Narratives

"Until the lion learns how to write, every story will glorify the hunter." -African Proverb.

For too long, Black and Brown communities have been the subjects of research—examined, dissected, and pathologized, rather than the architects of inquiry. Public health and social science have historically mined data from urban neighborhoods without returning insight, resources, or power to the communities from which that data was extracted. The result is a cycle of narrative dispossession: communities are denied the authority to name their conditions, define their priorities, or imagine their own futures. This chapter examines how that cycle is being disrupted, how data, once wielded as an instrument of marginalization, is increasingly being reclaimed as a tool of liberation.

From participatory action research in inner-city neighborhoods to community advisory boards guiding biomedical studies, marginalized people are reshaping the politics of inquiry. Rooted in lived experience, these models treat community members not as informants but as co-researchers, knowledge holders, and analysts. In places where institutions once ignored local wisdom, residents are now leading surveys, mapping health disparities, collecting oral histories, and driving program design. This shift does more than rebalance power; it produces research that is more relevant, accountable, and just. The reclamation of data is not simply about evidence—it is about self-determination.

The Data We Carry

Every data point carries a history. For Black and Brown men in the United States, that history is often one of surveillance, scrutiny, and silencing. In public health and social science, research instruments have long entered neighborhoods where trust was already fractured communities studied more than served. Whether documenting incarceration rates, mapping trauma exposure, or tracking chronic disease in urban centers, research on Black and Brown men has too often been conducted from a distance, without meaningful community input, shared ownership, or reciprocal benefit. The result is a data archive rich in statistics but poor in justice.

The extractive nature of traditional research is felt acutely by Black and Brown men, whose lives are frequently reduced to risk profiles and pathologies—criminality, absenteeism, violence, noncompliance rather than examined through lenses of resilience, leadership, or care. Contemporary reviews of urban health and criminal-legal research show that communities of color remain heavily studied while remaining underrepresented in research leadership and decision-making roles, particularly in studies involving policing, incarceration, and men's health.[1] Research on Black men, in particular, continues to prioritize deficit-based outcomes, disease prevalence, low healthcare utilization, educational disengagement, while rarely centering structural causation or community assets.[2]

These patterns are not accidental. They are shaped by the afterlives of historical abuses, most notably the Tuskegee Syphilis Study but also by more routine practices of racialized surveillance embedded in housing, education, healthcare, and criminal-legal systems. Contemporary scholarship shows that mistrust among Black men toward research institutions is not rooted in ignorance, but in lived experience with exploitation, misrepresentation, and the absence of accountability.[3] Data collection, in this context, has often functioned as an extension of control rather than care.

Yet data is not inherently oppressive. It can be reclaimed.For Black and Brown men, reclaiming research means reclaiming narrative authority. Participatory and community-engaged research models, particularly those involving formerly incarcerated men, LGBTQIA+ men, barbers, fathers, and neighborhood leaders allow participants to become co-investigators rather than extractive subjects. Studies rooted in community-based participatory research (CBPR) consistently demonstrate higher trust, improved data validity, and greater policy relevance when community members are involved in question formation, data interpretation, and dissemination.[4]

Initiatives such as barbershop-based health research, community-led men's health collaboratives, and participatory mapping projects have shown that when Black men help design and conduct research, the resulting data captures lived realities often erased by traditional methodologies, stress, dignity, care, and survival alongside disease and risk.[5] A recent systematic review found that CBPR approaches involving men of color were associated with stronger engagement, greater cultural resonance, and a higher likelihood of producing actionable change at the community or policy level.[6]

The data Black and Brown men carry is not only numerical—it is embodied. It lives in incarceration histories, stress-mediated illness, environmental exposure, and inherited trauma. But within that data also reside resistance, innovation, and care. When men reclaim the right to study their own lives, research shifts from a tool of surveillance to one of transformation. This chapter traces those shifts, how Black and Brown men are no longer only being studied, but are increasingly doing the studying, defining the terms, and healing the narrative.

The Politics of Being Studied

To be studied is not a neutral experience. For Black and Brown communities, research has rarely arrived as care; it has more often arrived as scrutiny. Clipboards, surveys, surveillance tools, and diagnostic frameworks have historically entered urban neighborhoods not to heal but to measure, categorize, and control. The question has rarely been *what do you need?* But rather *what's wrong with you?* This asymmetry defines the politics of being studied.

Public health and social science research have long positioned Black and Brown men as problems to be solved rather than partners in knowledge production. Studies disproportionately focus on risk—violence, disease prevalence, incarceration, substance use—while ignoring context, resilience, and structural causation. The result is a body of literature that documents harm without interrogating its origins. Data is collected downstream, after policy has failed, and then used to reinforce narratives of dysfunction rather than accountability.

This pattern reflects what many community members recognize instinctively: research has functioned as a form of extraction. Neighborhoods are mined for data in much the same way they have been mined for labor, votes, and cultural capital. Findings are published, grants renewed, careers advanced—while the communities that generated the data see little material benefit. Reports circulate in academic journals and policy briefings but rarely return to the streets they describe in ways that improve daily life. Knowledge flows upward; consequences remain local.

For Black and Brown men, this extractive logic is compounded by surveillance. Research is often indistinguishable from policing. Data collection mirrors carceral practices: observation without consent, categorization without context, intervention without trust. Health surveillance maps overlap neatly with maps of stop-and-frisk, probation density, environmental exposure, and school discipline. In this terrain, being studied feels less like participation and more like monitoring.

This history explains why mistrust of research is not cultural pathology but rational response. Skepticism is often framed as "hesitancy," "noncompliance," or "lack of health literacy," obscuring the structural reasons communities resist engagement. When research has repeatedly failed to protect, benefit, or even respect those it studies, refusal becomes a form of self-defense. Silence, withdrawal, and guarded participation are not barriers to research; they are data points themselves—signals of institutional failure.

The politics of being studied are also gendered. Masculinity norms that reward stoicism, emotional containment, and self-reliance shape how men engage with research. Many Black and Brown men have learned that disclosure carries risk: vulnerability can be weaponized, misinterpreted, or documented without care. As a result, men are often underrepresented in qualitative health research or present through flattened metrics that miss emotional complexity. What looks like disengagement is often strategic silence.

Importantly, harm does not require malicious intent. Even well-meaning studies can reproduce injustice when communities are positioned as objects rather than authors of inquiry. Research questions are often pre-formed before community engagement begins. Outcomes are defined by funders, not residents. Success is measured by publication rather than impact. In these conditions, "inclusion" becomes cosmetic, and participation becomes symbolic rather than transformative.

To understand the politics of being studied is to recognize that data is never neutral. They are shaped by power: who asks the questions, who defines the variables, who interprets the results, and who benefits from the answers. For Black and Brown men, the cost of being studied has too often been misrepresentation, pathologization, and erasure. Before research can heal, it must reckon with this history—not defensively, but honestly.

This recognition sets the stage for the next section. If harm has been institutional rather than incidental, then ethical repair must move beyond consent forms and review boards. The question is no longer whether research follows protocol, but whether it honors people. That reckoning begins with confronting institutional betrayal itself.

Institutional Betrayal and Ethical Reckonings

Ethics in research is often treated as a procedural hurdle rather than a moral commitment. Institutional Review Boards (IRBs), consent forms, and compliance training are designed to minimize harm, yet for many Black and Brown communities, harm has occurred within these very frameworks. This dissonance—between ethical approval and ethical practice reveals a deeper truth: institutional ethics have historically prioritized institutional protection over community well-being.

Institutional betrayal occurs when organizations tasked with safeguarding individuals instead reproduce harm, neglect, or exploitation. In the context of research, this betrayal is not always overt. It appears in studies that extract data without return, in projects that document trauma without offering support, and in publications that advance academic careers while communities remain unchanged. The question is not whether research follows the rules, but whether it honors responsibility.

For Black and Brown men in urban environments, this betrayal is cumulative. Research has tracked their bodies through incarceration records, hospital admissions, and surveillance datasets, while rarely asking how participation feels, what it costs, or what it yields. IRBs may approve a study because it poses "minimal risk," yet that calculation often ignores historical context. When mistrust is labeled as reluctance rather than wisdom, ethical review becomes detached from lived reality.

The legacy of unethical research—Tuskegee being the most cited but not the only example, casts a long shadow. Contemporary studies may no longer withhold treatment, but they often reproduce the same power imbalance: researchers observe, communities comply, and institutions benefit. Consent is obtained, but collaboration is absent. Data is collected, but healing is deferred. In this way, modern research can replicate the logic of past abuses without replicating their methods.

Ethical reckoning requires expanding the definition of harm. Harm is not only physical injury or privacy violation; it includes narrative distortion, emotional extraction, and the erosion of trust. It includes asking men to recount violence, loss, or incarceration without providing space for integration or care. It includes publishing deficit-focused findings that reinforce stigma while omitting resilience, creativity, and resistance. This reckoning also exposes the limits of neutrality. Claims of objectivity often mask unequal power: who frames the research question, who interprets the data, who authors the conclusions. When institutions position themselves as neutral arbiters of knowledge, they obscure the ways research participates in social ordering, deciding whose suffering is legible and whose expertise counts.

Importantly, institutional betrayal is not inevitable. It is produced through choices: what gets funded, whose voices are included, how success is measured. Ethical research demands more than procedural compliance; it demands relational accountability. It asks not only Can we study this? but Should we—and with whom? It requires institutions to reckon with their own role in producing harm and to accept that repair, not just regulation, is necessary.

This chapter does not argue for abandoning research, but for transforming it. Ethical reckoning is the bridge between critique and possibility. It opens the door to a different model—one where research is not something done to communities, but something built with them. The next section turns to that possibility: what happens when data is reclaimed, stories are honored, and knowledge becomes a site of healing rather than extraction.

Data Reclamation as Healing

Data reclamation is not merely a methodological shift; it is a reparative act. For communities that have long been measured, surveilled, and misrepresented, reclaiming data means reclaiming authorship over lived reality. It transforms research from an extractive enterprise into a process of collective meaning-making—one that restores dignity, agency, and trust. For Black and Brown men in urban environments, whose bodies and behaviors have been relentlessly quantified but rarely contextualized, data reclamation offers a pathway from being studied to being seen.

Participatory research methodologies such as Participatory Action Research (PAR) and Community-Based Participatory Research (CBPR) are central to this transformation. Unlike traditional top-down models, these approaches involve community members at every stage of inquiry from defining research questions to collecting, interpreting, and disseminating findings. The goal is not inclusion as a gesture, but power-sharing as a principle. Contemporary studies show that PAR and CBPR not only produce more culturally grounded data, but also improve trust, relevance, and sustainability of interventions in marginalized communities.[7] When Black and Brown men participate as co-researchers rather than subjects, research becomes a site of affirmation rather than surveillance.

Narrative-based methods play a particularly powerful role in this process. Oral history, photovoice, and community storytelling allow participants to articulate experience in their own language, rhythms, and frames of meaning. These methods are especially vital for Black and Brown men, whose emotional lives are often rendered invisible by both masculinity norms and deficit-oriented research frameworks. Photovoice, for example, enables participants to document environmental stressors, sources of pride, and everyday survival through images and collective reflection, shifting knowledge production from abstraction to embodiment.[8] Story becomes both data and medicine.

Reclaiming narrative authority also has measurable psychological and communal effects. Research indicates that participatory storytelling and co-interpretation foster empowerment, reduce internalized stigma, and strengthen collective efficacy, key determinants of mental and physical health.[9] For men whose lives have been shaped by incarceration, economic exclusion, or racialized policing, the opportunity to name one's own experience disrupts shame-based narratives and restores coherence. Being a narrator rather than an object of inquiry recalibrates how men see themselves and how they are seen by others.

At the community level, data reclamation builds durable infrastructure for change. When communities control data, they are more likely to mobilize findings for advocacy, policy reform, and resource allocation. Recent CBPR evaluations demonstrate that community-led data initiatives increase the likelihood that research findings are translated into concrete interventions, particularly in areas related to environmental health, chronic disease prevention, and violence reduction.[10] Ownership changes outcomes. Data becomes not an endpoint, but a tool for action.

Importantly, data reclamation also improves science itself. Community-driven research generates questions that institutions often overlook and interpretations that external researchers may miss. It challenges narrow definitions of rigor by insisting that validity includes cultural resonance, historical context, and lived expertise. In this way, reclamation does not reject science, it deepens it. It insists that knowledge grounded in relationship is more accurate than knowledge extracted at a distance.

For Black and Brown men, reclaiming data is a refusal to be reduced to risk profiles or pathology charts. It is an assertion that their lives contain insight, not just indicators. When communities reclaim research, data becomes a site of healing—repairing not only narratives, but the trust between bodies, knowledge, and possibility. The next section turns to what this reclamation looks like in practice, tracing how community-led methods are reshaping research on the ground.

Community-Led Methods in Practice

Translating theory into practice, community-led research initiatives across the United States are transforming what it means to generate, interpret, and protect data in ways that center Black and Brown lives. These efforts reject extractive research traditions and instead advance research justice, data sovereignty, and equitable knowledge production. From youth-led political organizing to grassroots data activism, community-centered inquiry demonstrates that research can function not as surveillance, but as survival strategy and, ultimately, as a tool of liberation.

Across these models, a common principle emerges those most impacted by inequality must have authority over how knowledge about their lives is produced, framed, and mobilized. Community-led research does not merely include residents as participants; it positions them as architects of inquiry, analysts of findings, and stewards of data. The following case studies illustrate how this shift unfolds on the ground.

Case Study: The Black Youth Project (BYP100) and Research Justice (Chicago, IL)

The Black Youth Project 100 (BYP100), founded in 2013, is a member-based organization of Black youth activists who engage in grassroots organizing, policy advocacy, and political education. Central to BYP100's ethos is the idea that research must serve the needs and visions of Black communities. Their work exemplifies "research justice"—a framework that asserts the right of marginalized communities to own, interpret, and use data to advocate for systemic change. For instance, their "Agenda to Build Black Futures" drew on survey data collected directly from Black youth, prioritizing the lived experiences and visions of participants. BYP100 integrates oral histories, participatory surveys, and storytelling into their research process to ensure that data collection affirms rather than exploits.[11] In internal evaluations and movement-based research produced by BYP100, participants report increased trust in the research process, higher levels of engagement, and stronger identification with project goals—key indicators of participatory relevance and retention.[12]

This model highlights the importance of community-led inquiry that not only seeks to understand community needs but also mobilizes data for organizing and policy change. From Chicago to Jackson, BYP100's model demonstrates how young people, particularly Black queer and trans youth, are using research to reclaim narratives, document harm, and push for liberation on their own terms.

Case Study: Data for Black Lives (D4BL) and Algorithmic Equity (Boston, MA)

Building on this model of grassroots data sovereignty, Data for Black Lives (D4BL) emerged in response to the racialized use of big data and predictive algorithms. Founded by Yeshimabeit Milner, D4BL brings together scientists, organizers, and technologists to interrogate how data systems impact Black communities. D4BL's mission is not just to critique harmful uses of data (such as predictive policing and biased credit algorithms) but to develop alternative tools that advance health equity, racial justice, and algorithmic accountability. One of their hallmark projects has involved community data audits—engaging Black communities in identifying and challenging local data practices that reinforce inequality. In Miami, D4BL led participatory tech audits to investigate the racial bias embedded in municipal surveillance tools and predictive analytics used in policing.[13]

Quantitative evaluations of D4BL's community data justice programs show that participants report significantly higher understanding of data ethics and systems.[14] Moreover, these programs have influenced policy debates around algorithmic fairness, particularly in areas like housing access and school discipline. Community participants in these projects report higher engagement and retention when their experiences are validated and when they co-create the questions guiding the data inquiry. D4BL's work demonstrates that algorithmic equity cannot be achieved without community control over how data are defined, collected, and applied.

While both BYP100 and D4BL focus on national and digital landscapes, local initiatives in cities like Baltimore and Detroit show how participatory research can be embedded in neighborhoods themselves, transforming relationships between institutions and residents. These models root research in place-based struggle, addressing the everyday material conditions that shape health and survival in Black and Brown communities.

Case Study: Baltimore's Community Data Lab and Detroit Urban Research Center (Baltimore, MD)

In Baltimore, the Community Data Lab (CDL) works at the intersection of public health, racial equity, and civic data transparency. Co-founded by researchers and local organizers, the CDL engages residents in defining and interpreting data about their own neighborhoods. Projects include mapping housing conditions, heat vulnerability, and access to food or healthcare services—done not just for residents, but with them. In a recent heat equity project, CDL trained high school students to gather air temperature and housing data using wearable sensors and surveys, revealing stark disparities in heat exposure between West Baltimore and wealthier neighborhoods.[15] Participants expressed increased trust in public data and reported that involvement helped them advocate for changes in zoning and green infrastructure.

Similarly, the Detroit Urban Research Center (URC) has been a national leader in community-based participatory research since 1995. Its projects involve longstanding partnerships between community-based organizations and academic institutions. One key initiative, "Healthy Environments Partnership," involved residents in conducting air quality monitoring, leading to local advocacy for reducing industrial pollution in majority-Black neighborhoods. Evaluation results showed that over 70% of participants reported stronger ties to local advocacy groups, and 65% reported using data from the project to engage in civic meetings or local government action.[16] The Detroit URC's work underscores how participatory evaluation centered on trust, relevance, and sustained collaboration can lead to real material benefits, including increased civic participation and improved public health policies.

Taken together, these case studies demonstrate that when communities, especially Black and Brown men and youth are equipped with research tools, they do more than collect data. They transform knowledge into strategies for survival, care, and resistance. Across local and national scales, community-led methods challenge the hierarchy of expertise that has long dominated public health and social science.

In these models, lived experience is not anecdotal, it is analytic. Storytelling is not supplementary, it is method. And research is no longer something done *to* communities, but something built *with* them. These practices insist on a fundamental truth: the knowledge required to heal communities already exists within them. The work of research justice is not to extract that knowledge, but to honor it, protect it, and put it back into the hands of those who live its consequences every day.

From Observation to Co-Creation

For decades, Black and Brown communities have been positioned as objects of observation rather than agents of knowledge production. Researchers entered neighborhoods with predetermined hypotheses, extracted stories and statistics, and left with publications, while communities were left unchanged or harmed. Even when framed as "inclusive," traditional research often preserved a hierarchy: experts asked the questions, communities supplied the data. Co-creation disrupts this logic. It shifts research from something done *to* communities into something built *with* them, redistributing power across the entire research process.

Co-creation is not simply participatory language layered onto conventional methods. It represents a fundamental epistemological shift. Communities help define research questions, determine what counts as evidence, interpret findings, and decide how and whether results are disseminated. This reorientation challenges dominant academic assumptions that equate rigor with distance and neutrality. Instead, rigor is reframed as relevance, accountability, and fidelity to lived experience.[17]

One of the clearest markers of this shift is authorship. In co-created research, community members are no longer confined to acknowledgments while academics claim intellectual ownership. Increasingly, participatory projects list community researchers, organizers, youth leaders, and formerly incarcerated men as co-authors recognizing that interpretation and contextual knowledge are as critical as statistical analysis. This practice directly challenges extractive norms within public health and social science, where communities generate insight, but institutions accrue prestige.[18]

Compensation further distinguishes co-creation from observation. Too often, marginalized communities are asked to contribute labor, trauma, and time under the guise of "engagement," while institutions receive funding and career advancement. Co-created research treats community expertise as labor worthy of pay, not charity. Ethical participatory frameworks now emphasize fair compensation, resource sharing, and budget transparency as core components of research justice.[19] When men are compensated for their intellectual and emotional contributions, research becomes less extractive and more sustainable.

Ownership of data is equally transformative. In traditional models, data belong to institutions stored in servers far removed from the neighborhoods that generated them. Co-creation insists on shared or community ownership through data-sharing agreements, local archives, and community-controlled dashboards. These practices ensure that findings remain accessible and actionable beyond the life of a grant, allowing communities to leverage data for advocacy, policy change, and self-determination.[20]

Co-creation also redefines outcomes. Academic success is often measured by publications and citations; community success looks different—policy shifts, improved services, collective healing, or narrative repair. Participatory evaluations show that community-defined outcomes are more likely to produce lasting health and social impact, precisely because they are grounded in local priorities rather than institutional metrics.[21]

For Black and Brown men in particular, co-creation addresses a long-standing methodological failure: the silencing of emotional and embodied knowledge. Men's health research has historically relied on deficit-based surveys that miss grief, fear, tenderness, and survival strategies. Co-created methods—listening sessions, peer-led interviews, narrative analysis—create conditions where men can speak on their own terms. When men help shape how questions are asked and where conversations occur, silence gives way to story, and data become more accurate as well as more humane.[2]

Ultimately, the shift from observation to co-creation does more than improve data quality; it repairs harm. It acknowledges that research has often functioned as surveillance and reimagines it as relationship. Co-created inquiry replaces extraction with shared accountability, transforming knowledge from institutional property into collective resource. In doing so, research moves beyond documenting inequality and begins to participate in its undoing.

Masculinity, Research, and Silence

Silence has long been misread as absence. In research on Black and Brown men, it is often interpreted as disengagement, resistance, or unreliability—a failure to participate fully in studies designed to measure health, emotion, and vulnerability. But silence is not emptiness. It is structure. It is learned. And it is methodological. Across public health, psychology, and sociology, men, particularly Black and Brown men are routinely described as "hard to reach" populations. Low survey completion rates, brief responses, or reluctance to disclose emotional distress are framed as cultural barriers or individual shortcomings. Rarely are these patterns interrogated as artifacts of research design itself. Masculine silence is treated as a problem to overcome, rather than a signal that the questions, settings, or power dynamics are misaligned.

Masculinity, as this book has shown, is not merely a social identity—it is a survival strategy forged under conditions of racialized exposure, economic precarity, surveillance, and bodily threat. For many men, emotional containment is not repression but protection. Silence becomes a learned response in environments where disclosure has historically been met with punishment, ridicule, medical neglect, or criminalization. When research ignores this context, it mistakes adaptation for pathology.

Traditional research instruments are particularly ill-equipped to capture this reality. Standardized surveys privilege verbal fluency, emotional labeling, and individual introspection skills that are unevenly rewarded across gendered and racialized contexts. Asking a man to rate his "emotional distress" on a Likert scale assumes not only psychological safety but also a cultural permission to name vulnerability. For men, whose bodies have been sites of surveillance—policed, incarcerated, medically experimented upon—silence can be a rational refusal rather than a deficit.

This failure of measurement has consequences. When men underreport stress, depression, fear, or grief, datasets reproduce the illusion that they are less affected by harm, less emotionally complex, or less in need of care. In reality, the costs surface elsewhere: in hypertension, metabolic disease, sleep disruption, substance use, aggression, and early mortality. What goes unspoken does not disappear, it relocates into the body. Silence becomes somatic. Earlier chapters traced how chronic stress, cortisol dysregulation, inflammation, and epigenetic changes accumulate in men living under constant threat. Chapter 5 showed how environmental pressures—heat, housing instability, endocrine disruption—shape behavior and physiology. Chapter 7 revealed how fatherhood and incarceration demand emotional suppression as a condition of survival. Together, these chapters expose a critical truth: men's silence is not evidence of resilience; it is often evidence of overload.

Research that fails to account for this dynamic risks compounding harm. When masculine silence is interpreted as noncompliance or lack of insight, men are further excluded from health interventions, funding priorities, and policy responses. Their suffering remains statistically invisible, reinforcing a cycle where absence of data is mistaken for absence of need. This is not neutrality, it is erasure. Importantly, silence is not universal. It is situational. When men are invited into spaces designed with trust, cultural resonance, and relational safety such as barbershops, peer circles, and community-led research teams, the data changes. Men speak differently. They narrate stress through story rather than symptom, through metaphor rather than metric. They describe grief as fatigue, fear as anger, love as protection. These are not distortions; they are alternative epistemologies.

Community-engaged research repeatedly demonstrates that when men help shape the research process, choosing the questions, methods, and settings disclosure increases and data quality improves. Silence loosens not because men are "opened up," but because the research relationship shifts from extraction to recognition. Masculinity, in these contexts, is not an obstacle to knowledge production; it is a source of insight into how harm is endured and how care is negotiated.

This reframing demands a methodological reckoning. Rather than asking why men do not speak, researchers must ask: *What makes speaking unsafe? What kinds of knowledge have we been trained to value? Whose ways of knowing have we dismissed as illegible?* Silence, when properly understood, is data. It marks histories of betrayal, survival, and constraint. It tells us where institutions have failed to earn trust. To study Black and Brown men without accounting for masculine silence is to misunderstand both masculinity and health. Emotional restraint, guardedness, and refusal are not evidence of disengagement, they are responses to worlds that have made vulnerability costly. Research that treats silence as pathology replicates the very power dynamics it claims to study.

Reimagining research, then, requires more than inclusion. It requires listening differently. It requires methods that honor embodiment, narrative, gesture, and absence. It requires recognizing that what is not said may be as informative as what is spoken—and that silence itself may be a form of testimony. In this light, masculinity is not a barrier to research, it is a diagnostic lens. It reveals where systems have demanded endurance over expression, control over care, survival over truth. When research learns to hear masculine silence not as void but as signal, it moves closer to justice. And when men are finally allowed to speak or not speak on their own terms, knowledge production becomes not an act of surveillance, but one of repair.

Reimaging Research as Resistance

If research has historically functioned as a tool of surveillance, extraction, and control, it can also be reimagined as a tool of resistance. To reclaim research is to refuse neutrality, to reject the fiction that knowledge production is detached from power. For Black and Brown communities, particularly men whose bodies have been sites of measurement but not meaning research becomes radical when it shifts from observing harm to interrupting it.

Reimagining research as resistance begins with a fundamental question: *Who benefits from the data?* Traditional academic models have privileged institutions over communities, publication over impact, and objectivity over accountability. In this framework, communities are mined for information while remaining excluded from interpretation, authorship, and ownership. Research justice demands a reversal of this flow. Data must move back toward the people whose lives generate it—not as charity, but as right.

When communities define research agendas, knowledge becomes actionable. Counter-mapping exposes environmental racism where official maps deny it. Community health audits document lived realities that administrative datasets obscure. Oral histories surface survival strategies that never appear in regression models. In these acts, research is no longer descriptive, it is insurgent. It names harm where institutions minimize it and demands repair where systems offer explanation instead of change.

This reimagining also requires abandoning the myth of detached observation. Research is never neutral. The choice of variables, the framing of questions, and the interpretation of silence all reflect values. To pretend otherwise is to obscure responsibility. Research as resistance insists on positionality, transparency, and ethical alignment. It recognizes that scholars are not outside the communities they study, and that accountability does not end with informed consent.

Importantly, resistance-oriented research does not reject rigor, it redefines it. Rigor is not only statistical power or methodological purity; it is relevance, trustworthiness, and consequence. A study that is methodologically elegant but socially inert fails the communities it claims to serve. By contrast, research that is co-created, culturally grounded, and mobilized for change embodies a different standard of excellence, one rooted in justice rather than prestige.

For Black and Brown men, reclaiming research is also about reclaiming narrative authority. It is the refusal to be reduced to risk profiles, crime statistics, or disease prevalence rates. It is the insistence that masculinity, fatherhood, silence, care, and survival are not confounders to be controlled for, but realities to be understood. When men participate as researchers, analysts, and storytellers, the data shift. Pathology gives way to context. Deficit gives way to depth.

This book itself participates in that resistance. It refuses to separate biology from policy, masculinity from metabolism, fatherhood from structural violence. It treats bodies as archives, neighborhoods as laboratories, and care as data. It does not ask Black and Brown men to explain themselves to systems that have already failed them. Instead, it interrogates the systems and demands that research do the same.

Reimagining research as resistance is ultimately about sovereignty. Knowledge sovereignty. Narrative sovereignty. Bodily sovereignty. It is about deciding that communities do not need to be saved by data but armed with it. When research aligns with liberation rather than legitimacy, it stops asking whether communities are resilient enough and starts asking why they have been forced to be.

The question, then, is no longer whether research can change the world. It is whether we are willing to let it.

Conclusion: Who Tells the Story?

Research justice is health justice. This is not a metaphor, nor a rhetorical flourish, it is a material truth for Black and Brown men whose bodies, behaviors, and communities have long been studied without being understood, measured without being respected, and cited without being served. For generations, health research has operated as a gatekeeping system, deciding not only what counts as evidence, but whose lives are legible within it. The result has been an epistemological harm layered atop biological and social injury: communities rendered visible only through deficit, risk, and pathology.

To ask *who tells the story* is to ask who holds power. What gets studied, how questions are framed, which outcomes are prioritized, and how findings are interpreted all shape the interventions that follow. For Black and Brown men in marginalized urban environments, these decisions are not abstract, they determine access to care, allocation of resources, and public narratives about worth and responsibility. When research excludes lived experience, it reproduces inequality. When it centers it, new possibilities emerge.

Research that heals does not begin with extraction. It begins with relationship. It recognizes that knowledge production is not a neutral process but a relational one, grounded in trust, accountability, and reciprocity. Healing-centered research acknowledges historical harm, resists voyeurism, and refuses to treat communities as laboratories rather than partners. For men whose bodies have been over-policed and under-cared for, research becomes restorative when it affirms agency, dignity, and voice.

Reclaiming research is therefore an act of resistance. It is the refusal to allow data to be used against the people from whom it is drawn. It is the insistence that Black and Brown men are not merely subjects of inquiry, but theorists of their own lives, holders of embodied knowledge about stress, survival, care, masculinity, and healing. When men participate as co-researchers, analysts, and narrators, the archive changes. Silence becomes signal. Context replaces caricature. Complexity displaces blame.

This shift carries consequences beyond academia. Research shapes policy. Policy shapes environments. Environments shape biology. When communities control knowledge production, health interventions become more relevant, more trusted, and more effective. Data stops justifying harm and starts demanding repair. In this way, research justice becomes a mechanism for collective transformation, not simply improved representation.

To support community knowledge production is not optional—it is an ethical mandate. It requires redistributing resources, rethinking authorship, compensating lived expertise, and dismantling institutional barriers that keep marginalized voices at the margins of scholarship. It asks researchers to relinquish control in favor of collaboration, and institutions to value impact over prestige.

Ultimately, this book argues that storytelling is not ancillary to health, it is foundational. Bodies remember what systems deny. Communities carry truths that datasets overlook. And when Black and Brown men are given the space to tell their own stories, on their own terms, research becomes what it was always meant to be: a tool not for observation alone, but for justice.

The question is no longer whether communities are ready to be heard.

The question is whether institutions are ready to listen.

Epilogue: Street Medicine, Future Maps

Street Medicine, Future Maps— Visions for a Healing-Centered World

> "We are not the problem. The problem is the systems that do not account for us, the systems that refuse to see our fullness. The future of health justice is not about fixing people; it's about fixing the world that harms them." -Amina Mohammed.

To imagine a future grounded in health equity, we must begin by reimagining what health itself means in marginalized communities. Equity cannot be an afterthought or a policy add-on; it must be the foundation upon which systems of care, support, and accountability are built. In this future, the voices of Black and Brown men are not peripheral but central—not consulted after decisions are made but shaping them from the start. Healing is no longer imposed from above, but cultivated from within communities through culturally responsive, healing-centered care. Grassroots movements, street medicine, and community advocacy redefine public health not as control, but as connection.

This vision requires honesty about how we arrived here. Health disparities among Black and Brown men did not emerge in a vacuum; they are the cumulative result of centuries of structural violence, racial exclusion, and gendered expectation. Masculinity, shaped by the demand to be strong, stoic, and self-reliant has often functioned as both shield and burden. While it offered protection in hostile environments, it also constrained vulnerability, delayed care-seeking, and buried emotional pain beneath survival. The body has carried the cost of that silence in elevated blood pressure, suppressed immunity, and shortened lives.

These pressures are intensified for gay and queer Black and Brown men, whose identities are often rendered illegible within rigid masculine norms and underserved by health systems structured around heteronormativity. Double marginalization by race and sexuality produces distinct health vulnerabilities, from mental health distress to disproportionate HIV burden. Yet the root remains the same: systems that punish difference, pathologize vulnerability, and deny care unless men conform. Any future committed to equity must therefore be expansive enough to hold the full range of masculinities, identities, and ways of being.

The future of health justice depends on reimagining masculinity itself—not as emotional containment, but as emotional fluency, not as dominance, but as stewardship. Masculinity must become a resource for healing rather than a barrier to it. This shift is not abstract. It is enacted every time a man seeks care without shame, mentors a young person, shows tenderness in public, or names his own pain as worthy of attention.

Street medicine offers a blueprint for this future. By bringing care directly into neighborhoods, onto sidewalks, into shelters, community centers, and encampments it rejects the notion that health belongs only to institutions. Street medicine meets people where they are, both geographically and emotionally. It recognizes that trust, dignity, and safety are as essential to healing as prescriptions or procedures. For Black and Brown men navigating geographic isolation, economic precarity, and medical mistrust, this model restores care as a relationship rather than a transaction.

Ultimately, healing cannot be an individual burden. It must be collective, relational, and structural. A healing-centered world acknowledges historical harm without being bound by it, and invests in communities not as sites of risk, but as sites of wisdom. This is the work of tending what happens *under the skin,* the stress responses, immune burdens, and inherited wounds while transforming what exists *above the pavement*: the streets, institutions, and systems that shape daily survival. In such a future, Black and Brown men are not passive recipients of intervention, but co-creators of health, knowledge, and possibility. By amplifying community voices, supporting grassroots infrastructure, and embracing care as a form of justice, we begin to map a world where biology and policy no longer work against each other—where healing is not imagined, but lived.

ABOUT THE AUTHOR

Carlin Dexter Nelson, PhD, MPH, CPH, CHES®, is an epidemiologist, public health educator and scholar committed to examining how systems of inequality shape health across urban communities. He is Certified in Public Health (CPH) and is a Certified Health Education Specialist (CHES), combining rigorous epidemiological methods with culturally grounded health promotion. He holds an Associate of Science (AAS) from Trident Technical College (TTC), a Bachelor of Arts (BA) in Public Health with a minor in Sociology from the College of Charleston (CofC), a Master of Public Health (MPH) in Epidemiology from the Medical University of South Carolina (MUSC), where he was an Interprofessional Education Fellow, and a Doctorate of Philosophy (PhD) in Public Health-Epidemiology from Walden University.

Currently, Dr. Nelson teaches in the Department of Health Sciences at Coppin State University in Baltimore, Maryland, where he leads courses in epidemiology, health statistics and research, health promotion, drug education, and urban health. His previous professional experience includes roles at the Drug Enforcement Administration (DEA), Truth Initiative, Johns Hopkins University, the Medical University of South Carolina (MUSC), College of Charleston, as well as a translational research fellow at Nemours Children's Hospital. His research has been published in the *American Journal of Preventive Medicine*, the *Journal of Racial and Ethnic Health Disparities*, the *International Journal of Human Rights in Healthcare*, the *Association of Black Nursing Faculty Foundation Journal (ABNFFJ)*, and the *Social Science Research Network (SSRN)*. Across these outlets, his work examines social determinants of health (SDOH), racial and gender disparities, and community-driven public health strategies.

A native of Charleston, South Carolina, Dr. Nelson grew up in Title I schools and a subsidized, single-parent household with experiences that fuel his passion for community-rooted scholarship. A proud TRiO Upward Bound alum who later returned as a mentor and administrator, has spent over five years mentoring first-generation and underserved students through TRiO and the NIH STEP-UP program. Dr. Nelson is driven by the belief that our present circumstances do not define our future. Through teaching, research, and writing, he continues to show up for the next generation of public health leaders, reminding them that someone, somewhere, at some point in time is counting on them to rise.

Notes
Chapter 1: The Biological Grind— Chronic Stress, Cortisol, Toxic Stress

1. Ekaterina Pesheva. "ZIP Code or Genetic Code?" Home, January 14, 2019. https://hms.harvard.edu/news/zip-code-or-genetic-code.
2. Centers for Disease Control and Prevention. (2023, September 6). *Social determinants of health: Know what affects health*. U.S.
3. Steven Booske et al., "Different Perspectives for Assigning Weights to Determinants of Health," Milbank Quarterly 88, no. 4 (2010): 7–14; David McGinnis, Pamela Williams-Russo, and James Knickman, "The Case for More Active Policy Attention to Health Promotion," Health Affairs 21, no. 2 (2002): 78–93.
4. David R. Williams and Selina A. Mohammed, "Racism and Health I: Pathways and Scientific Evidence," *American Behavioral Scientist* 57, no. 8 (2013): 1152–1173, https://doi.org/10.1177/0002764213487340.
5. American Psychological Association, *Stress in America 2022: Concerned for the Future, Beset by Inflation*, Stress in America Survey (Washington, DC: American Psychological Association, 2022), https://www.apa.org/news/press/releases/stress/2022/report-octobe
6. McEwen, Bruce S., and Harold Akil. "Revisiting the Stress Concept." *Journal of Neuroscience* 40, no. 1 (2020): 12–21. Bruce S. McEwen and Eliot Stellar, "Stress and the Individual: Mechanisms Leading to Disease," Archives of Internal Medicine 153, no. 18 (1993): 2093–2101.
7. McEwen, Bruce S., and Eliot Stellar. "Stress and the Individual." *Archives of Internal Medicine* 153, no. 18 (1993): 2093–2101.
8. Teresa E. Seeman et al., "Socio-Economic Differentials in Peripheral Biology: Cumulative Allostatic Load," *Annals of the New York Academy of Sciences* 1186 (2010): 223–239.
9. Richard Rothstein, *The Color of Law: A Forgotten History of How Our Government Segregated America* (New York: Liveright, 2017).
10. Geronimus, Arline T., and John P. Thompson. "Racial Inequality in Health and Policy-Induced Crises." *Du Bois Review* 17, no. 2 (2020): 267–287.
11. Elizabeth H. Blackburn and Elissa S. Epel, *The Telomere Effect: A Revolutionary Approach to Living Younger, Healthier, Longer* (New York: Grand Central Publishing, 2017).
12. Elissa S. Epel et al., "Accelerated Telomere Shortening in Response to Life Stress," *Proceedings of the National Academy of Sciences* 101, no. 49 (2004): 17312–17315; Needham, Belinda L., et al. "Neighborhood Characteristics and Telomere Length." *Health & Place* 31 (2015): 209–216.

13. Arline T. Geronimus, "Understanding and Eliminating Racial Inequalities in Women's Health in the United States: The Role of the Weathering Conceptual Framework," *Journal of the American Medical Women's Association* 61, no. 2 (2006): 133–136.
14. Mayo Clinic Staff. (2023, March 23). *Cortisol: What it does and why it matters.* Mayo Clinic. https://www.mayoclinic.org/healthy-lifestyle/stress-management/in-depth/cortisol/art-20046033
15. Guidi, J., Lucente, M., Sonino, N., & Fava, G. A. (2021). Allostatic load and its impact on health: A systematic review. Psychotherapy and Psychosomatics, 90(1), 11–27. https://doi.org/10.1159/000515820; McEwen, Bruce S., and Eliot Stellar. "Stress and the Individual." *Archives of Internal Medicine* 153, no. 18 (1993): 2093–2101.; McEwen, B. S., & Seeman, T. (1999). Protective and damaging effects of mediators of stress: Elaborating and testing the concepts of allostasis and allostatic load. *Annals of the New York Academy of Sciences, 896*(1), 30–47. https://doi.org/10.1111/j.1749-6632.1999.tb08103.x
16. Hackett, Robert A., and Andrew Steptoe. "Type 2 Diabetes Mellitus and Psychological Stress—A Modifiable Risk Factor." *Nature Reviews Endocrinology* 13, no. 9 (2017): 547–560. https://doi.org/10.1038/nrendo.2017.64
17. Wulsin, Lawrence R., and Bruce M. Singal. "Do Depressive Symptoms Increase the Risk for the Onset of Coronary Disease? A Systematic Quantitative Review." *Psychosomatic Medicine* 65, no. 2 (2003): 201–210. https://doi.org/10.1097/01.PSY.0000058371.50240.E3.; Leong, Shiao-Yng, Woon-Man Tan, and Weng-Tin Tan. "Chronic Stress, Cortisol Dysregulation, and Risk of Type 2 Diabetes." *Journal of Clinical Endocrinology & Metabolism* 105, no. 8 (2020): e2871–e2880. https://doi.org/10.1210/clinem/dgaa351
18. Slavich, George M., and Michael R. Irwin. "From Stress to Inflammation and Major Depressive Disorder: A Social Signal Transduction Theory of Depression." *Psychological Bulletin* 140, no. 3 (2014): 774–815. https://doi.org/10.1037/a0035302.; Furman, David, Judith Campisi, Eric Verdin, et al. "Chronic Inflammation in the Etiology of Disease across the Life Span." *Nature Medicine* 25, no. 12 (2019): 1822–1832. https://doi.org/10.1038/s41591-019-0675-0.
19. Lupien, Sonia J., Bruce S. McEwen, Megan R. Gunnar, and Christine Heim. "Effects of Stress throughout the Lifespan on the Brain, Behaviour and Cognition." *Nature Reviews Neuroscience* 10, no. 6 (2018): 434–445. https://doi.org/10.1038/nrn2639.
20. Stawski, Robert S., et al. "Associations between Cortisol and Cognitive Aging: Longitudinal Evidence from Midlife to Older Adulthood." *Proceedings of the National Academy of Sciences of the United States of America* 120, no. 6 (2023): e2216824120. https://doi.org/10.1073/pnas.2216824120.
21. Tavares, C. D., Bell, C. N., Zare, H., Hudson, D., & Thorpe, R. J. (2022). Allostatic load, income, and race among Black and White men in the United

States. *American Journal of Men's Health, 16*(2), 15579883221092290. https://doi.org/10.1177/15579883221092290

22. Smith, Matthew L., et al. "Racial Differences in Stress Biomarkers and Allostatic Load among U.S. Men." *Journal of Men's Health* 17, no. 3 (2021): 45–55.; Tavares, Ana I., et al. "Chronic Stress, Cortisol Dysregulation, and Health Inequities among Urban Black Men." *Social Science & Medicine* 296 (2022): 114756. https://doi.org/10.1016/j.socscimed.2022.114756.

23. Nelson, R. H., Connolly, N. D. B., & Fishback, P. V. (2018). Mapping inequality: Redlining in New Deal America. *University of Richmond*. https://dsl.richmond.edu/panorama/redlining/; Rothstein, R. (2017). *The color of law: A forgotten history of how our government segregated America*. Liveright Publishing Corporation.

24. Nelson, C. D. (2023). *An association of the transgenerational implications of redlining and obesity on pediatric type II diabetes* (Doctoral dissertation). Walden University. https://scholarworks.waldenu.edu/dissertations/12485/

25. American Lung Association. (2023). *State of the Air 2023*. https://www.lung.org/getmedia/338b0c3c-6bf8-480f-9e6e-b93868c6c476/SOTA-2023.pdf

26. Hoffman, J. S., Shandas, V., & Pendleton, N. (2020). The effects of historical housing policies on resident exposure to intra-urban heat: A study of 108 US urban areas. *Nature Communications, 11*(1), 1–10. https://doi.org/10.1038/s41467-020-18364-1

27. Stansfeld, S. A., Clark, C., & Alfred, T. (2018). Environmental noise and health: A review of the evidence. *Environmental Health Perspectives, 126*(11), 116001. https://doi.org/10.1289/EHP.1805

28. Centers for Disease Control and Prevention. (2023). *Lead poisoning prevention: Protecting children from lead exposure*. Centers for Disease Control and Prevention. https://www.cdc.gov/nceh/lead/default.htm

29. The Lancet. (2022). The impact of police violence on Black Americans: A public health crisis. *The Lancet, 400*(10346), 1881-1892. https://doi.org/10.1016/S0140-6736(22)01407-7

30. Kershaw, K. N., et al. (2023). Neighborhood racial segregation linked to shorter life spans. *JAMA Health Forum, 4*(7), e2312465. https://doi.org/10.1001/jamahealthforum.2023.12465

31. Geronimus, A. T., Hicken, M. T., Keene, D., & Bound, J. (2006). "Weathering" and age patterns of allostatic load scores among Blacks and Whites in the United States. *American Journal of Public Health, 96*(5), 826–833. https://doi.org/10.2105/AJPH.2004.060749

32. Courtenay, W. H. (2000). Constructions of masculinity and their influence on men's well-being: A theory of gender and health. *Social Science & Medicine, 50*(10), 1385–1401. https://doi.org/10.1016/S0277-9536(99)00390-1; Jordan, H., Jeremiah, R., Watson, K., Corte, C., Steffen, A., & Matthews, A. K. (2024).

Exploring preventive health care utilization among Black/African American men. *American Journal of Men's Health, 18*(1), 15579883231225548. https://doi.org/10.1177/15579883231225548

33. Smith, J. T., Williams, A. L., & Jones, C. M. (2021). Masculine norms and mental health outcomes among Black men. *American Journal of Men's Health, 15*(4), 55-62. https://doi.org/10.1177/1557988321997213
34. Centers for Disease Control and Prevention (CDC). (2020). *Heart disease and stroke risk among Black men.* https://www.cdc.gov/heartdisease/facts.htm
35. Semenza, D., Baker, N., & Smith, A. (2024, May 17). Gun violence touches nearly 60 percent of Black Americans – and predicts disability. *Rutgers University.* https://ritms.rutgers.edu/news/gun-violence-touches-nearly-60-percent-of-black-americans-and-predicts-disability/
36. Lanfear, C. C., Bucci, R., Kirk, D., & Sampson, R. J. (2023). Inequalities in exposure to firearm violence by race, sex, and birth cohort from childhood to age 40 years, 1995–2021. *JAMA Network Open, 6*(5), e2312465. https://doi.org/10.1001/jamanetworkopen.2023.12465

Chapter 2: Masculinity as a Risk Factor— Norms, Suppression, and Survival

1. Levant, R. F., Wimer, D. J., & Williams, C. M. (2014). Masculinity constructs as protective buffers and risk factors for men's health. *American Journal of Men's Health, 8*(2), 110–120.; Pan American Health Organization. (2018). Addressing masculinity and men's health to advance universal health and gender equality. *Revista Panamericana de Salud Pública, 42*, e196.
2. Wong, Y. J., & Rochlen, A. B. (2018). Masculinity, emotion regulation, and psychopathology: A critical review and integrated model. *Clinical Psychology Review, 66*, 106–116.
3. Levant, R. F., Wimer, D. J., & Williams, C. M. (2014). Masculinity constructs as protective buffers and risk factors for men's health. *American Journal of Men's Health, 8*(2), 110–120.
4. Addis, M. E., & Mahalik, J. R. (2003). Men, masculinity, and the contexts of help seeking. *American Psychologist, 58*(1), 5–14.
5. FDA Voices. (2025, June). *The concerning trend in men's health.* U.S. Food & Drug Administration.; Pan American Health Organization. (2019, November 18). *1 in 5 men will not reach the age of 50 in the Americas, due to issues relating to toxic masculinity.* PAHO.; Addis, M. E., & Mahalik, J. R. (2003). Men, masculinity, and the contexts of help seeking. *American Psychologist, 58*(1), 5–14.; Walther, A., Eggenberger, N., ... (2024). Masculine ideals double suicide risk in men. *Neuroscience News.*
6. Connell, R. W., & Messerschmidt, J. W. (2005). Hegemonic masculinity: Rethinking the concept. *Gender & Society, 19*(6), 829–859. https://doi.org/10.1177/0891243205278639

7. Boxer, A., & Gill, P. R. (2021). Predicting anxiety from the complex interaction between masculinity and spiritual beliefs. *American Journal of Men's Health, 15*(5), 15579883211049021.
8. Pan American Health Organization. (2018). Addressing masculinity and men's health to advance universal health and gender equality. *Revista Panamericana de Salud Pública, 42*, e196.
9. James, Sherman A. 1994. "John Henryism and the Health of African Americans." *Culture, Medicine and Psychiatry* 18 (2): 163–182. https://doi.org/10.1007/BF01309983
10. Geronimus, A. T., Hicken, M. T., Keene, D., & Bound, J. (2006). "Weathering" and age patterns of allostatic load scores among Blacks and Whites in the United States. *American Journal of Public Health, 96*(5), 826–833. https://doi.org/10.2105/AJPH.2004.060749
11. American Heart Association. 2025. *Heart Disease and Stroke Statistics—2025 Update: A Report from the American Heart Association.* Dallas, TX: American Heart Association.
12. Centers for Disease Control and Prevention (CDC). 2010. *National Ambulatory Medical Care Survey: Summary Tables.* Atlanta, GA: U.S. Department of Health and Human Services.; Centers for Disease Control and Prevention (CDC). 2023. *Health, United States, 2023.* Atlanta, GA: U.S. Department of Health and Human Services.; Axios. 2024. "Why Men Avoid the Doctor." *Axios*, April 2024. https://www.axios.com.
    1. Landis, Adelie. "The Hidden Hurt, Men's Mental Health." Elite DNA Behavioral Health, August 1, 2025. https://elitedna.com/the-hidden-hurt-mens-mental-health/#:~:text=Mental%20health%20issues%20in%20men,realize%20they%20are%20feeling%20depressed.
13. National Health Interview Survey (NHIS). 2021. *Mental Health Treatment Utilization Tables.* Hyattsville, MD: National Center for Health Statistics.
14. Courtenay, Will H. 2000. "Constructions of Masculinity and Their Influence on Men's Well-Being: A Theory of Gender and Health." *Social Science & Medicine* 50 (10): 1385–1401. https://doi.org/10.1016/S0277-9536(99)00390-1; Curtin, S. C., Brown, K. A., & Jordan, M. E. (2022, November). *Suicide rates for the three leading methods by race and ethnicity: United States, 2000–2020* (NCHS Data Brief No. 450). National Center for Health Statistics. https://www.cdc.gov/nchs/products/databriefs/db450.htm
15. American Cancer Society. "Black Men and Prostate Cancer." American Cancer Society Cancer Action Network, November 5, 2025. https://www.fightcancer.org/policy-resources/black-men-and-prostate-cancer-0.
16. Majors, R., & Billson, J. M. (1993). *Cool pose: The dilemmas of Black manhood in America.* Touchstone.
17. American Psychological Association. (2024, February 20). *APA Foundation launches new initiative for mental health of Black men and boys.*

https://www.apaf.org/media-events/news/apa-foundation-launches-new-initiative-for-mental-health-of-black-men-and-boys-5903fe6b0ef2f5c4529e0ed93c3caf06/
18. The Sentencing Project. (2021, October 13). *New report finds imprisonment rate of Black men has fallen by nearly 50% since 2000, but pushback threatens continued progress.* https://www.sentencingproject.org/press-releases/new-report-finds-imprisonment-rate-of-black-men-has-fallen-by-nearly-50-since-2000-but-pushb American Civil Liberties Union. (n.d.). *Mass incarceration.* https://www.aclu.org/news/smart-justice/mass-incarceration; Prison Policy Initiative. (n.d.). *Racial and ethnic disparities in incarceration.* https://www.prisonpolicy.org/research/racial_and_ethnic_disparities/
19. Couloute, L., & Kopf, D. (2018). *Out of prison & out of work: Unemployment among formerly incarcerated people.* Prison Policy Initiative. https://www.prisonpolicy.org/reports/outofwork.html; American Bar Association. (2021). *Overlooked and undervalued: Ex-offenders in the employment market.* American Bar Association Journal of Labor & Employment Law, 37(1). https://www.americanbar.org/content/dam/aba/publications/aba_journal_of_labor_employment_law/v37/no-1/jlel-37-1-6.pdf
20. American Civil Liberties Union. (n.d.). *With AI and criminal justice, the devil is in the data.* ACLU. https://www.aclu.org/news/criminal-law-reform/with-ai-and-criminal-justice-the-devil-is-in-the-data
21. Curtin, S. C., Brown, K. A., & Jordan, M. E. (2022, November). *Suicide rates for the three leading methods by race and ethnicity: United States, 2000–2020* (NCHS Data Brief No. 450). National Center for Health Statistics. https://www.cdc.gov/nchs/products/databriefs/db450.htm
22. Epsy, W., Allen, A. M., Jones, C. A., & Smith, L. M. (2023). Achieving mental health equity in Black male suicide prevention. *Journal of Racial and Ethnic Health Disparities.* https://doi.org/10.1007/s40615-023-01520-2
23. Springer, K. W., & Mouzon, D. (2011). "Macho men" and preventive health care: Implications for older men in different social classes. *Journal of Health and Social Behavior, 52*(2), 212–227. https://doi.org/10.1177/0022146510393972; Sileo, K. M., Kershaw, T. S., et al. (2020). Dimensions of masculine norms, depression, and mental health service utilization: Results from a prospective cohort study among emerging adult men in the United States. *American Journal of Men's Health, 14*(3), 1557988320906980. https://doi.org/10.1177/1557988320906980; Vogel, D. L., Heimerdinger-Edwards, S. R., Hammer, J. H., & Hubbard, A. (2011). *Psychological Help-Seeking Stigma and Men's Attitudes Toward Professional Help. Journal of Counseling Psychology, 58*(2), 274–281. https://doi.org/10.1037/a002225
24. Santo, Loredana, Zachary J Peters, Lello Guluma, and Jill J. Ashman. National Health Statistics reports number 211 n October 22, 2024. Accessed December 20, 2025. https://www.cdc.gov/nchs/data/nhsr/nhsr211.pdf.

25. Centers for Disease Control and Prevention. (2013, July 22). *Racial differences in life expectancy*. National Center for Health Statistics. https://blogs.cdc.gov/nchs/2013/07/22/1651/; National Institutes of Health. (2017, October 23). *Study: Heart disease, stroke cutting lives of Black Americans*. National Heart, Lung, and Blood Institute. https://www.nhlbi.nih.gov/news/2017/study-heart-disease-stroke-cutting-lives-black-americans
26. Waters, H., & Graf, M. (2018, May). *The cost of chronic diseases in the U.S.* Milken Institute. https://milkeninstitute.org/sites/default/files/reports-pdf/Chronic-Disease-Executive-Summary-r2.pdf

Chapter 3: Masked in Plain Sight—Queer Masculinities and the Politics of Passing

1. Human Rights Campaign Foundation. (2015). *Black and African American LGBTQ youth report*. https://www.hrc.org/resources/black-and-african-american-lgbtq-youth-report
2. Van der Star, A., Bränström, R., & Pachankis, J. E. (2019). Sexual orientation openness and depression symptoms: A population-based study. *Psychology of Sexual Orientation and Gender Diversity, 6*(2), 199–208. https://doi.org/10.1037/sgd0000311
3. Pascoe, C. J. (2005). 'Dude, you're a fag': Adolescent masculinity and the fag discourse. *Sexualities, 8*(3), 329–346. https://doi.org/10.1177/1363460705053337
4. Oshana, M. E., McCabe, C. J., & Blashill, A. J. (2020). Body dysmorphic disorder symptoms among sexual minority men: The role of minority stressors. *Psychology of Sexual Orientation and Gender Diversity, 7*(2), 230–237. https://doi.org/10.1037/sgd0000355
5. The Trevor Project. Research brief: LGBTQ youth and body dissatisfaction, January 2023. https://www.thetrevorproject.org/wp-content/uploads/2023/01/January_2023_Research_Brief_Final.pdf.
6. Nowicki, G. P., Marchwinski, B. R., O'Flynn, J. L., Griffiths, S., & Rodgers, R. F. (2022). Body image and associated factors among sexual minority men: A systematic review. Body Image, 43, 154–169. https://doi.org/10.1016/j.bodyim.2022.08.006
7. University of Brighton. (2022, February 17). *Plus-size men face stigma in gay spaces*. https://www.brighton.ac.uk/news/2022/new-study-shows-plus-size-men-face-stigma-in-gay-spaces
8. QX Magazine. (2022, February 21). *Plus-size men face stigma in gay spaces*. https://www.qxmagazine.com/2022/02/plus-size-men-face-stigma-in-gay-spaces

9. Bränström, R., & Pachankis, J. E. (2020). Sexual orientation concealment and depression symptoms: A population-based study. Psychology of Sexual Orientation and Gender Diversity, 6(2), 199–208.
10. Pachankis, J. E., & Bränström, R. (2015). Sexual orientation concealment and mental health: A conceptual and meta-analytic review. Psychological Bulletin, 141(5), 1228–1264.; Bränström, R., & Pachankis, J. E. (2020). Sexual orientation concealment and depression symptoms: A population-based study. Psychology of Sexual Orientation and Gender Diversity, 6(2), 199–208.; Williams Institute. (2021, January 23). Black LGBTQ+ adults experience severe health barriers. Them.us. https://www.them.us/story/black-lgbtq-health-issues; Hitch, A. E., & Brown, J. L. (2023). Sexual orientation concealment and mental health among BIPOC men who have sex with men: A scoping review. Journal of Gay & Lesbian Mental Health, 28(3), 1–24.
11. Pew Research Center. (2021, February 16). *Faith among Black Americans.* https://www.pewresearch.org/religion/2021/02/16/faith-among-black-americans/
12. Centers for Disease Control and Prevention. (2021). *HIV among African American gay and bisexual men.* https://stacks.cdc.gov/view/cdc/26487
13. Meyer IH: Prejudice, social stress, and mental health in lesbian, gay, and bisexual populations. Psychol Bull 2003, 129:674–697. First integrative review, meta-analysis, and integrative articulation of minority stress theory as an explanation for mental health inequalities faced by sexual minority individuals.
14. Turpin, R., Thorpe, R. J., Jr., & Gaskins, S. (2021). Psychometric validation of the Connectedness to the LGBT Community Scale among Black sexual minority men living with HIV. *LGBT Health, 8*(2), 125–132. https://doi.org/10.1089/lgbt.2020.0242

Chapter 4: Hood Epigenetics—Trauma, Memory, and Molecular Legacy

1. Geronimus, A. T. (1992). The weathering hypothesis and the health of African American women and men. Ethnicity & Disease, 2(3), 207–221.; Gravlee, C. C. (2009). How race becomes biology: Embodiment of social inequality. American Journal of Physical Anthropology, 139(1), 47–57.; Krieger, N. (2014). Discrimination and health inequities. International Journal of Health Services, 44(4), 643–710. https://doi.org/10.2190/HS.44.4.b
2. Brody, G. H., Yu, T., Chen, E., Beach, S. R., & Miller, G. E. Supportive Family Environments Ameliorate the Link Between Racial Discrimination and Epigenetic Aging: A Replication Across Two Longitudinal Cohorts. Psychol Sci. 2016 Apr;27(4):530-41. doi: 10.1177/0956797615626703. Epub 2016 Feb 25. PMID: 26917213; PMCID: PMC4833531.
3. Geronimus, A. T. (1992). The weathering hypothesis and the health of African American women and men. Ethnicity & Disease, 2(3), 207–221.

4. Yehuda, R., Daskalakis, N. P., & et al. (2016). Childhood maltreatment and vulnerability to PTSD: The role of epigenetic pathways. Biological Psychiatry, 79(7), 1-9. https://doi.org/10.1016/j.biopsych.2015.11.029; Parade, S. H., et al. (2016). Child maltreatment and telomere shortening. Development and Psychopathology, 28(4pt2), 1487–1495.
5. McGowan, P. O., Sasaki, A., D'Alessio, A. C., & et al. (2009). Epigenetic regulation of the glucocorticoid receptor in human brain associates with childhood abuse. Nature Neuroscience, 12(3), 342-348. https://doi.org/10.1038/nn.2270.
6. Yehuda, R., Daskalakis, N. P., & et al. (2016). Childhood maltreatment and vulnerability to PTSD: The role of epigenetic pathways. Biological Psychiatry, 79(7), 1-9. https://doi.org/10.1016/j.biopsych.2015.11.029
7. Kuzawa, C. W., & Sweet, E. (2009). Epigenetic pathways to disease. The American Journal of Human Biology, 21(5), 505-510. https://doi.org/10.1002/ajhb.20977
8. Turecki, G., & Meaney, M. J. (2016). Effects of the social environment and stress on glucocorticoid receptor gene methylation: Implications for the developmental origins of health and disease. American Journal of Psychiatry, 173(7), 645-657. https://doi.org/10.1176/appi.ajp.2016.16030343
9. Kipling, D., & Cooke, H. J. (1990). Telomeres and telomerase: Their significance in ageing and cancer. Nature, 344(6263), 589-591. https://doi.org/10.1038/344589a0
10. Epel, E. S., Blackburn, E. H., Lin, J., Dhabhar, F. S., Adler, N. E., Morrow, J. D., & Cawthon, R. M. (2004). Accelerated telomere shortening in response to life stress. Proceedings of the National Academy of Sciences, 101(49), 17312-17315. https://doi.org/10.1073/pnas.0407162101
11. Chae, D. H., Lincoln, K. D., Adler, N. E., & Syme, S. L. (2014). The association between racial discrimination and telomere length in African Americans. American Journal of Epidemiology, 179(2), 345-354. https://doi.org/10.1093/aje/kwt230
12. Black, D. S., & Slavich, G. M. (2016). Mindfulness meditation and the immune system: A systematic review of randomized controlled trials. *Annals of the New York Academy of Sciences, 1373*(1), 13–24. https://doi.org/10.1111/nyas.12998
13. Gross, J. J., & John, O. P. (2019). Individual differences in two emotion regulation processes: Implications for affect, relationships, and well-being. Journal of Personality and Social Psychology, 85(2), 311–325. https://doi.org/10.1037/0022-3514.85.2.311
14. Kiecolt-Glaser, J. K., McGuire, L., Robles, T. F., & Glaser, R. (2013). Chapter 3: The impact of stress on the immune system. In S. A. P. J. E. D. de Souza (Ed.), Handbook of Stress and the Brain (pp. 63–87). Elsevier. https://doi.org/10.1016/B978-0-12-417083-7.00003-0
15. Simon, L., & Admon, R. (2023). From childhood adversity to latent stress vulnerability in adulthood: the mediating roles of sleep disturbances and HPA

axis dysfunction. *Neuropsychopharmacology: official publication of the American College of Neuropsychopharmacology, 48*(10), 1425–1435. https://doi.org/10.1038/s41386-023-01638-9

16. Rahim Kurwa (2020): The New MantintheHouse Rules: How the Regulation of Housing Vouchers Turns Personal Bonds into Eviction Liabilities, Housing Policy Debate, DOI: 10.1080/10511482.2020.1778056
17. Kaliman, P., Alvarez-López, M. J., Cosín-Tomás, M., Rosenkranz, M. A., Lutz, A., & Davidson, R. J. (2014). Rapid changes in histone deacetylases and inflammatory gene expression in expert meditators. *Psychoneuroendocrinology, 40*, 96–107. https://doi.org/10.1016/j.psyneuen.2013.11.004
18. Schutte, N. S., & Malouff, J. M. (2014). A meta-analytic review of the effects of mindfulness meditation on telomerase activity. *Psychoneuroendocrinology, 42*, 45–48. https://doi.org/10.1016/j.psyneuen.2013.12.017
19. The Confess Project. (2021). *America's first mental health barbershop movement.* https://www.theconfessproject.com/
20. Roots Community Health Center. (2021). *The Emancipator Initiative: Empowering communities through holistic care.* https://rootsclinic.org/emancipator-initiative/
21. Meaney, M. J. (2010). Epigenetics and the biological definition of gene × environment interactions. *Child Development, 81*(1), 41–79. https://doi.org/10.1111/j.1467-8624.2009.01381.x; Nestler, E. J., Peña, C. J., Kundakovic, M., Mitchell, A., & Akbarian, S. (2016). Epigenetic basis of mental illness. *The Neuroscientist, 22*(5), 447–463. https://doi.org/10.1177/1073858415608147
22. Bower, J. E., Irwin, M. R., & Cole, S. W. (2014). Mind–body interventions and immune system outcomes: A meta-analysis. *Brain, Behavior, and Immunity, 37*, 1–16. https://doi.org/10.1016/j.bbi.2013.10.010; Kaliman, P., Alvarez-López, M. J., Cosín-Tomás, M., Rosenkranz, M. A., Lutz, A., & Davidson, R. J. (2014). Rapid changes in histone deacetylases and inflammatory gene expression in expert meditators. *Psychoneuroendocrinology, 40*, 96–107. https://doi.org/10.1016/j.psyneuen.2013.11.004
23. Pennebaker, J. W., & Smyth, J. M. (2016). *Opening up by writing it down: How expressive writing improves health and eases emotional pain* (3rd ed.). Guilford Press.
24. Chen, E., Miller, G. E., Kobor, M. S., & Cole, S. W. (2011). Maternal warmth buffers the effects of low early-life socioeconomic status on pro-inflammatory signaling in adulthood. *Molecular Psychiatry, 16*(7), 729–737. https://doi.org/10.1038/mp.2010.53

Chapter 5: Heat, Hormones, and Housing: Environmental Endocrinology in the Inner City

EPILOGUE: STREET MEDICINE, FUTURE MAPS 163

1. Hoffman, J. S., Shandas, V., & Pendleton, N. (2020). The effects of historical housing policies on redlining and urban heat. Environmental Health Perspectives, 128(1), 017006.
2. Basu, R., Dominici, F., & Samet, J. M. (2020). The effects of extreme heat on human mortality: A multi-city case study in the United States. *Environmental Health Perspectives, 128*(7), 77001. https://doi.org/10.1289/EHP7176.
3. Shea, M., McCann, S., & Gorman, S. (2016). Impact of chronic heat exposure on cortisol levels and metabolic health. *International Journal of Environmental Health Research, 26*(1), 40-50. https://doi.org/10.1080/09603123.2015.1089534
4. Radwan, M., Jebari, S., & Behbehani, K. (2018). The impact of environmental stressors on male reproductive health: A study on the effects of temperature, pollution, and exposure to toxic chemicals. *Journal of Environmental Health, 81*(9), 1-9. https://doi.org/10.1080/00207233.2018.1537452
5. Radwan, E. M., Ashraf, M., & Ali, W. A. (2018). Chronic stress, environmental pollutants, and testosterone levels in Black men: Impact on health and life satisfaction. Journal of Men's Health, 15(2), 75-84. https://doi.org/10.1016/j.jomh.2017.10.008
6. Hoffman, J. S., Shandas, V., & Pendleton, N. (2020). The effects of historical housing policies on resident exposure to intra-urban heat: A study of 108 US urban areas. *Nature Communications, 11*(1), 1–10. https://doi.org/10.1038/s41467-020-18364-1
7. "About Lead in Paint." Centers for Disease Control and Prevention. https://www.cdc.gov/lead-prevention/prevention/paint.html.; Environmental Protection Agency (EPA). (2020). *Integrated Science Assessment (ISA) for Particulate Matter (Final Report)*. https://www.epa.gov/isa/integrated-science-assessments-isa
8. Environmental Protection Agency (EPA). (2021). *EPA report on hazardous waste sites and environmental health disparities.* https://www.epa.gov/environmentaljustice; Environmental Protection Agency. (2020). *Air Quality Statistics Report.* https://www.epa.gov/air-trends/air-quality-national-summary
9. U.S. Census Bureau. (2019). American housing survey: 2019. U.S. Department of Commerce, Economics and Statistics Administration. https://www.census.gov/programs-surveys/ahs.html
10. Habtemichael, E.N., Li, D.T., Camporez, J.P. *et al.* Insulin-stimulated endoproteolytic TUG cleavage links energy expenditure with glucose uptake. *Nat Metab* 3, 378–393 (2021). https://doi.org/10.1038/s42255-021-00359-x; Ko S. H. (2024). Effects of Heat Stress-Induced Sex Hormone Dysregulation on Reproduction and Growth in Male Adolescents and Beneficial Foods. *Nutrients, 16*(17), 3032. https://doi.org/10.3390/nu16173032; Leproult, R., & Van Cauter, E. (2011). Effect of 1 week of sleep restriction on testosterone levels in young healthy men. *JAMA, 305*(21), 2173–2174. https://doi.org/10.1001/jama.2011.710

11. National Low Income Housing Coalition. (2020). *The gap: A shortage of affordable homes*. National Low Income Housing Coalition. https://nlihc.org/gap
12. Ashe, N., Wozniak, S., Conner, M., Ahmed, R., Demetres, M. R., Makarem, N., Tehranifar, P., Nandakumar, R., & Ghosh, A. (2023). Association of extreme heat events with sleep and cardiovascular health: A scoping review. *Research Square*, rs.3.rs-3678410. https://doi.org/10.21203/rs.3.rs-3678410/v1; Habtemichael, E.N., Li, D.T., Camporez, J.P. et al. Insulin-stimulated endoproteolytic TUG cleavage links energy expenditure with glucose uptake. *Nat Metab* 3, 378–393 (2021). https://doi.org/10.1038/s42255-021-00359-x; Ko S. H. (2024). Effects of Heat Stress-Induced Sex Hormone Dysregulation on Reproduction and Growth in Male Adolescents and Beneficial Foods. *Nutrients*, 16(17), 3032. https://doi.org/10.3390/nu16173032; Leproult, R., & Van Cauter, E. (2011). Effect of 1 week of sleep restriction on testosterone levels in young healthy men. *JAMA*, 305(21), 2173–2174. https://doi.org/10.1001/jama.2011.710
13. Patel, R. P., Brooks, C. L., & Herrera, C. G. (2021). Environmental stressors and cardiovascular risk in marginalized communities: The role of heat, pollution, and housing. American Journal of Public Health, 111(3), 411-419. https://doi.org/10.2105/AJPH.2020.305785
14. Gustafson, T. E., Dubois, A., & St. John, A. M. (2020). Environmental stress and cardiovascular health: The role of the autonomic nervous system. International Journal of Environmental Health Research, 30(5), 592-603. https://doi.org/10.1080/09603123.2020.1775309
15. Harlan, S. L., Chow, J., & An, L. (2006). A climate for change: The role of climate change in shaping urban health. Urban Climate, 7, 33–48. https://doi.org/10.1016/j.uclim.2020.100347
16. Cohen, S., Janicki-Deverts, D., & Miller, G. E. (2020). Psychological stress and disease. JAMA, 298(14), 1685–1687. https://doi.org/10.1001/jama.298.14.1685
17. Bagley, Kathleen. "High Stress Can Cause Testosterone to Drop." SynergenX Health, June 12, 2024. https://synergenxhealth.com/low-testosterone-treatment-high-stress-can-cause-testosterone-drop/#:~:text=Many%20men%20also%20experience%20a,activities%20that%20require
18. Shea, M., McCann, S., & Gorman, S. (2016). Impact of chronic heat exposure on cortisol levels and metabolic health. *International Journal of Environmental Health Research*, 26(1), 40-50. https://doi.org/10.1080/09603123.2015.1089534; Shea, K. M., Lipinski, P., & Young, J. T. (2016). Cortisol dysregulation from environmental stressors: Impact on immune function and metabolic health in marginalized populations. Environmental Health Perspectives, 124(6), 840-846. https://doi.org/10.1289/ehp.1509617
19. Sapolsky, R. M. (2004). *Why zebras don't get ulcers: The acclaimed guide to stress, stress-related diseases, and coping* (3rd ed.). Holt Paperbacks.

20. Mahalik, J. R., et al. (2003). Masculinity and perceived normative health behaviors in men: The moderating role of gender role orientation. Journal of Social and Clinical Psychology, 22(3), 277-289. https://doi.org/10.1521/jscp.22.3.277.21862
21. Santos, A., et al. (2017). Masculinity and emotional responses to environmental stressors: The role of cultural norms. Psychology of Men & Masculinities, 18(4), 508-516. https://doi.org/10.1037/men0000078
22. Anderson, C. A. (2001). Heat and violence. Current Directions in Psychological Science, 10(1), 33-38. https://doi.org/10.1111/1467-8721.00109; Carleton, T. (2017). Impact of climate change on social violence. Climatic Change, 142(3-4), 485-499. https://doi.org/10.1007/s10584-017-1975-3
23. Dunn, L., et al. (2020). Cognitive and emotional processing during heat waves: Implications for aggression in low-income urban environments. Environmental Health Perspectives, 128(12), 127003. https://doi.org/10.1289/EHP7187
24. Cohn, A., et al. (2018). Emotional regulation and aggression in response to heat stress in marginalized communities. Psychological Science, 29(5), 734-745. https://doi.org/10.1177/0956797618768766
25. Zhao, Z., et al. (2019). Urban heat and violent crime: Evidence from U.S. cities. Journal of Environmental Economics and Management, 98, 1-19. https://doi.org/10.1016/j.jeem.2019.01.004
26. Zavaleta, J. R., et al. (2015). Heat exposure and violent behavior: A study in marginalized urban populations. Journal of Urban Health, 92(5), 781-791. https://doi.org/10.1007/s11524-015-9999-7
27. Hoffman, J. S., Shandas, V., & Pendleton, N. (2020). The effects of historical housing policies on resident exposure to intra-urban heat: A study of 108 US urban areas. Nature Communications, 11(1), 1–10. https://doi.org/10.1038/s41467-020-18364-1
28. Voelkel, J., Hellman, D., Sakuma, R., & Shandas, V. (2018). Assessing vulnerability to urban heat: A study of disproportionate heat exposure and access to refuge by socio-demographic status in Portland, Oregon. International Journal of Environmental Research and Public Health, 15(4), 640. https://doi.org/10.3390/ijerph15040640
29. Swope, C. B., & Hernández, D. (2019). Housing as a determinant of health equity: A conceptual model. Social Science & Medicine, 243, 112571. https://doi.org/10.1016/j.socscimed.2019.112571
30. Yoon, S., Katz, L., Nadeem, E., & Seidman, E. (2021). Heat vulnerability and housing insecurity in New York City: A cross-sectional study of environmental health disparities. Journal of Urban Health, 98(3), 480–492. https://doi.org/10.1007/s11524-021-00526-2
31. Locke, D. H., Hall, B., Grove, J. M., Pickett, S. T., Ogden, L. A., Aoki, C., ... & O'Neil-Dunne, J. P. (2021). Residential housing segregation and urban

tree canopy in 37 U.S. cities. NPJ Urban Sustainability, 1, Article 15. https://doi.org/10.1038/s42949-021-00015-w
32. Wilhelmi, O. V., & Hayden, M. H. (2010). Connecting people and place: A new framework for reducing urban vulnerability to extreme heat. Environmental Research Letters, 5(1), 014021. https://doi.org/10.1088/1748-9326/5/1/014021
33. Anguelovski, Isabelle, James J. T. Connolly, Hamil Pearsall, and J. Timmons Roberts. "Why Green 'Climate Gentrification' Threatens Poor and Vulnerable Populations." PNAS. https://www.pnas.org/doi/pdf/10.1073/pnas.1920490117.
34. Keenan, J. M., Hill, T., & Gumber, A. (2018). Climate gentrification: From theory to empiricism in Miami-Dade County, Florida. Environmental Research Letters, 13(5), 054001. https://doi.org/10.1088/1748-9326/aabb32
35. Taylor, D. E. (2020). Environmental justice: Establishing new spaces for equity and inclusion in climate change discourse. Nature Climate Change, 10(9), 784–790. https://doi.org/10.1038/s41558-020-0826-3
36. Hondula, D. M., Balling, R. C., Vanos, J. K., & Georgescu, M. (2019). Rising temperatures, human health, and the role of adaptation. Current Climate Change Reports, 5(1), 1–10. https://doi.org/10.1007/s40641-019-00121-4
37. City of Baltimore. (2022). Cooling cities strategy: Advancing climate resilience through equity-centered heat mitigation. Baltimore Office of Sustainability. https://www.baltimoresustainability.org/plans/cooling-cities-strategy
38. Bullard, R. D. (2000). Dumping in Dixie: Race, class, and environmental quality (3rd ed.). Westview Press.
39. Bullard, R. D. (2000). Dumping in Dixie: Race, class, and environmental quality (3rd ed.). Westview Press.; Taylor, D. E. (2020). Environmental justice: Establishing new spaces for equity and inclusion in climate change discourse. Nature Climate Change, 10(9), 784–790. https://doi.org/10.1038/s41558-020-0826-3
40. McEwen, B. S. (2007). Physiology and neurobiology of stress and adaptation: Central role of the brain. Physiological Reviews, 87(3), 873–904. https://doi.org/10.1152/physrev.00041.2006; Gee, G. C., & Payne-Sturges, D. C. (2004). Environmental health disparities: A framework integrating psychosocial and environmental concepts. Environmental Health Perspectives, 112(17), 1645–1653. https://doi.org/10.1289/ehp.7074
41. Geronimus, A. T., Hicken, M. T., Keene, D., & Bound, J. (2006). "Weathering" and age patterns of allostatic load scores among Blacks and Whites in the United States. American Journal of Public Health, 96(5), 826–833. https://doi.org/10.2105/AJPH.2004.060749

42. Richardson, L., Norris, F. H., & Joseph, J. (2021). Community trauma and the toxic stress response: Implications for Black men. Journal of Urban Health, 98(1), 15–25. https://doi.org/10.1007/s11524-021-00529-3
43. Swinburn, B. A., Kraak, V. I., Allender, S., et al. (2019). The global syndemic of obesity, undernutrition, and climate change: The Lancet Commission report. The Lancet, 393(10173), 791–846. https://doi.org/10.1016/S0140-6736(18)32822-8

Chapter 6: Urban Immunity — Food Deserts, Inflammation, and Resistance

1. Geronimus, A. T., Hicken, M. T., Keene, D., & Bound, J. (2006). "Weathering" and age patterns of allostatic load scores among Blacks and Whites in the United States. American Journal of Public Health, 96(5), 826–833. https://doi.org/10.2105/AJPH.2004.060749; Libby, P. (2002). Inflammation in atherosclerosis. Nature, 420(6917), 868–874. https://doi.org/10.1038/nature01323; Thayer, J. F., & Sternberg, E. M. (2010). Beyond heart rate variability: Vagal regulation of allostatic systems. Annals of the New York Academy of Sciences, 1186(1), 129–141. https://doi.org/10.1111/j.1749-6632.2009.05378.x
2. Odoms-Young, A., & Bruce, M. A. (2018). Examining the Impact of Structural Racism on Food Insecurity: Implications for Addressing Racial/Ethnic Disparities. Family & community health, 41 Suppl 2 Suppl, Food Insecurity and Obesity (Suppl 2 FOOD INSECURITY AND OBESITY), S3–S6. https://doi.org/10.1097/FCH.0000000000000183
3. Bradley, Katharine & Herrera, Hank. (2015). Decolonizing Food Justice: Naming, Resisting, and Researching Colonizing Forces in the Movement. Antipode. 48. 10.1111/anti.12165.
4. Shaker, Y., Grineski, S. E., Collins, T. W., & Flores, A. B. (2023). Redlining, racism and food access in US urban cores. Agriculture and human values, 40(1), 101–112. https://doi.org/10.1007/s10460-022-10340-3
5. Huang H. A Spatial Analysis of Obesity: Interaction of Urban Food Environments and Racial Segregation in Chicago. J Urban Health. 2021 Oct;98(5):676-686. doi: 10.1007/s11524-021-00553-y. Epub 2021 Jul 15. PMID: 34264475; PMCID: PMC8280681.; Zenk, S. N., Schulz, A. J., Israel, B. A., James, S. A., Bao, S., & Wilson, M. L. (2005). Neighborhood racial composition, neighborhood poverty, and the spatial accessibility of supermarkets in metropolitan Detroit. American journal of public health, 95(4), 660–667. https://doi.org/10.2105/AJPH.2004.042150
6. Benavides-Colón, Amelia. "Food - Social Determinants of Health in Detroit." Social Determinants of Health in Detroit - Social Determinants of Health in Detroit, February 20, 2025. https://sdoh.planetdetroit.org/food/.
7. Cooksey-Stowers, K., Schwartz, M. B., & Brownell, K. D. (2017). Food Swamps Predict Obesity Rates Better Than Food Deserts in the United

States. *International journal of environmental research and public health, 14*(11), 1366. https://doi.org/10.3390/ijerph14111366
8. Welsh, A., Hammad, M., Piña, I. L., & Kulinski, J. (2024). Obesity and cardiovascular health. *European journal of preventive cardiology, 31*(8), 1026–1035. https://doi.org/10.1093/eurjpc/zwae025; Kyalwazi, A. N., Loccoh, E. C., Brewer, L. C., Ofili, E. O., Xu, J., Song, Y., Joynt Maddox, K. E., Yeh, R. W., & Wadhera, R. K. (2022). Disparities in Cardiovascular Mortality Between Black and White Adults in the United States, 1999 to 2019. *Circulation, 146*(3), 211–228. https://doi.org/10.1161/CIRCULATIONAHA.122.060199
9. Baltimore City 2024 Food Environment Brief, February 6, 2025. https://planning.baltimorecity.gov/baltimore-food-policy-initiative/food-environment#:~:text=%E2%80%9CFood%20desert%E2%80%9D%20suggests%20there%20is%2C
10. Hilmers, A., Hilmers, D. C., & Dave, J. (2012). Neighborhood disparities in access to healthy foods and their effects on environmental justice. *American journal of public health, 102*(9), 1644–1654. https://doi.org/10.2105/AJPH.2012.300865; 1.; Petroka K, Campbell-Bussiere R, Dychtwald DK, Milliron B-J. Barriers and facilitators to healthy eating and disease self-management among older adults residing in subsidized housing. *Nutrition and Health.* 2017;23(3):167-175. doi:10.1177/0260106017722724; Lee, Wonhyung & Jurkowski, Janine & Gentile, Nicole. (2023). Food Pantries and Food Deserts: Health Implications of Access to Emergency Food in Low-Income Neighborhoods. Urban Social Work. 7. 29-42. 10.1891/USW-2022-0008.
11. Libby, P. (2002). Inflammation in atherosclerosis. *Nature, 420*(6917), 868–874. https://doi.org/10.1038/nature01323; Thayer, J. F., & Sternberg, E. M. (2010). Beyond heart rate variability: Vagal regulation of allostatic systems. *Annals of the New York Academy of Sciences, 1186*(1), 129–141. https://doi.org/10.1111/j.1749-6632.2009.05378.x
12. Environmental Protection Agency. (2020). *Air Quality Statistics Report.* https://www.epa.gov/air-trends/air-quality-national-summary
13. Nishimura, K. K., Galanter, J. M., Roth, L. A., Oh, S. S., Thakur, N., Nguyen, E. A., Thyne, S., Farber, H. J., Serebrisky, D., Kumar, R., Brigino-Buenaventura, E., Davis, A., LeNoir, M. A., Meade, K., Rodriguez-Cintron, W., Avila, P. C., Borrell, L. N., Bibbins-Domingo, K., Rodriguez-Santana, J. R., Sen, Ś., ... Burchard, E. G. (2013). Early-life air pollution and asthma risk in minority children. The GALA II and SAGE II studies. *American journal of respiratory and critical care medicine, 188*(3), 309–318. https://doi.org/10.1164/rccm.201302-0264OC
14. Mortimer, K. M., Neas, L. M., Dockery, D. W., Redline, S., & Tager, I. B. (2002). The effect of air pollution on inner-city children with asthma. *The European respiratory journal, 19*(4), 699–705. https://doi.org/10.1183/09031936.02.00247102

15. "'Asthma Alley': Why Minorities Bear Burden of Pollution Inequity Caused by White People." The Guardian, April 4, 2019. https://www.theguardian.com/us-news/2019/apr/04/new-york-south-bronx-minorities-pollution-inequity.
16. Alvarez C. H. (2023). Structural Racism as an Environmental Justice Issue: A Multilevel Analysis of the State Racism Index and Environmental Health Risk from Air Toxics. *Journal of racial and ethnic health disparities, 10*(1), 244–258. https://doi.org/10.1007/s40615-021-01215-0
17. Ryan, P. H., Zanobetti, A., Coull, B. A., Andrews, H., Bacharier, L. B., Bailey, D., Beamer, P. I., Blossom, J., Brokamp, C., Datta, S., Hartert, T., Khurana Hershey, G. K., Jackson, D. J., Johnson, C. C., Joseph, C., Kahn, J., Lothrop, N., Louisias, M., Luttmann-Gibson, H., Martinez, F. D., … Gold, D. R. (2024). The Legacy of Redlining: Increasing Childhood Asthma Disparities through Neighborhood Poverty. *American journal of respiratory and critical care medicine, 210*(10), 1201–1209. https://doi.org/10.1164/rccm.202309-1702OC
18. Slopen, N., Non, A. L., Williams, D. R., Roberts, A. L., & Albert, M. A. (2016). Childhood adversity, adult neighborhood context, and cumulative biological risk for chronic diseases in adulthood. *Psychosomatic Medicine, 78*(4), 492–501. https://doi.org/10.1097/PSY.0000000000000301
19. Geronimus, A. T., Hicken, M. T., Keene, D., & Bound, J. (2006). "Weathering" and age patterns of allostatic load scores among Blacks and Whites in the United States. *American Journal of Public Health, 96*(5), 826–833. https://doi.org/10.2105/AJPH.2004.060749
20. Thayer, J. F., & Sternberg, E. M. (2010). Beyond heart rate variability: Vagal regulation of allostatic systems. *Annals of the New York Academy of Sciences,* 1186(1), 129–141. https://doi.org/10.1111/j.1749-6632.2009.05378.x; Thayer, J. F., & Sternberg, E. M. (2010). Neural aspects of immunomodulation: Focus on the vagus nerve. *Brain, Behavior, and Immunity, 24*(8), 1223–1228. https://doi.org/10.1016/j.bbi.2010.07.247
21. Jackson, J. S., Knight, K. M., & Rafferty, J. A. (2010). Race and unhealthy behaviors: Chronic stress, the HPA axis, and physical and mental health disparities over the life course. *American Journal of Public Health,* 100(5), 933–939. https://doi.org/10.2105/AJPH.2008.143446
22. Berger, M., & Sarnyai, Z. (2015). More than skin deep: Stress neurobiology and mental health consequences of racial discrimination. *Stress,* 18(1), 1–10. https://doi.org/10.3109/10253890.2014.989204
23. Courtenay, W. H. (2000). Constructions of masculinity and their influence on men's well-being: A theory of gender and health. *Social Science & Medicine,* 50(10), 1385–1401. https://doi.org/10.1016/S0277-9536(99)00390-1
24. Springer, K. W., & Mouzon, D. (2011). "Macho men" and preventive health care: Implications for older men in different social classes. *Journal of Health and Social Behavior, 52*(2), 212–227. https://doi.org/10.1177/0022146510393972

25. Centers for Disease Control and Prevention (CDC). (2022). Life expectancy by race and sex, United States. National Center for Health Statistics.
26. Foucault, M. (1978). The history of sexuality: Volume 1 (R. Hurley, Trans.). Pantheon Books.
27. Mirajkar, A., Oswald, A., Rivera, M., Logan, G., Macintosh, T., Walker, A., Lebowitz, D., & Ganti, L. (2023). Racial disparities in patients hospitalized for COVID-19. *Journal of the National Medical Association, 115*(4), 436–440. https://doi.org/10.1016/j.jnma.2023.06.006; Jefferson, C., Watson, E., Certa, J. M., Gordon, K. S., Park, L. S., D'Souza, G., Benning, L., Abraham, A. G., Agil, D., Napravnik, S., Silverberg, M. J., Leyden, W. A., Skarbinski, J., Williams, C., Althoff, K. N., Horberg, M. A., & NA-ACCORD Corona-Infectious-Virus Epidemiology Team (CIVET) (2022). Differences in COVID-19 testing and adverse outcomes by race, ethnicity, sex, and health system setting in a large diverse US cohort. *PloS one, 17*(11), e0276742. https://doi.org/10.1371/journal.pone.0276742
28. Bullard, R. D., Mohai, P., Saha, R., & Wright, B. (2007). Toxic wastes and race at twenty. Environmental Law, 38, 371–411.
29. Krieger, N. (2014). Discrimination and health inequities. International Journal of Health Services, 44(4), 643–710. https://doi.org/10.2190/HS.44.4.b
30. Berger, M., & Sarnyai, Z. (2015). More than skin deep: Stress neurobiology and mental health consequences of racial discrimination. *Stress, 18*(1), 1–10. https://doi.org/10.3109/10253890.2014.989204; Courtenay, W. H. (2000). Constructions of masculinity and their influence on men's well-being: A theory of gender and health. *Social Science & Medicine, 50*(10), 1385–1401. https://doi.org/10.1016/S0277-9536(99)00390-1
31. Ford, C. L., Airhihenbuwa, C. O., & Walker, D. (2019). The role of research in addressing health inequities: The case of Black men. *Journal of Men's Health, 15*(1), 22–31. https://doi.org/10.1016/j.jomh.2018.11.007
32. Liu, C., Liu, L., & Wang, Y. (2017). Impact of air pollution on sperm motility in urban Chinese men: A longitudinal analysis. Environmental Science & Technology, 51(10), 5672-5681. https://doi.org/10.1021/acs.est.7b01622
33. Li, D., Xie, Y., & Wang, Z. (2021). The effect of urban air pollution on sperm quality: A cross-sectional study in China. Environmental Health Perspectives, 129(5), 057001. https://doi.org/10.1289/EHP7255
34. White, M. M. (2018). Freedom farmers: Agricultural resistance and the Black freedom movement. University of North Carolina Press.
35. Algert, S. J., Baameur, A., & Renvall, M. J. (2016). Vegetable output and cost savings of community gardens in San Jose, California. Journal of the Academy of Nutrition and Dietetics, 116(1), 107–113. https://doi.org/10.1016/j.jand.2015.07.007
36. Detroit Black Community Food Security Network (DBCFSN). (2021). D-Town Farm: Urban agriculture as resistance. https://www.dbcfsn.org/

37. ChiFresh Kitchen. (2022). Worker-owned cooperative building food justice in Chicago. https://www.chifreshkitchen.com/
38. Solnit, R., & Fong, T. (2023). The pandemic as portal: Mutual aid, social change, and community care in COVID-19. Haymarket Books.
39. Gilmore, R. W. (2022). Abolition geography: Essays towards liberation (B. Gilmore, Ed.). Verso Books.
40. Acta Non Verba (ANV Farm). (2023). About us. https://anvfarm.org/about/
41. Grow Greater Englewood (GGE). (2021). Food access and youth empowerment through agriculture. https://www.growgreaterenglewood.org/
42. Johns Hopkins Center for a Livable Future. (2022). Baltimore Food Policy Initiative and FoodRx pilot program. https://clf.jhsph.edu/projects/baltimore-food-policy-initiative
43. Hager, E. R., Cockerham, A., O'Reilly, N., Harrington, D., Harding, J., Hurley, K. M., & Black, M. M. (2017). Food insecurity and nutrition program participation among low-income preschool-aged children in Baltimore City. Journal of Nutrition Education and Behavior, 49(7), 636–645. https://doi.org/10.1016/j.jneb.2017.04.004
44. Finley, R. (2013). A guerilla gardener in South Central LA [TED Talk]. TED Conferences. https://www.ted.com/talks/ron_finley_a_guerilla_gardener_in_south_central_la

Chapter 7: Concrete Wombs and Metal Cradles—Fatherhood, Care, and Inherited Wounds

1. Jones, J., Smith, L., & Caldwell, C. (2023). *Profiles of father involvement among unmarried Black fathers and associations with child socio-emotional outcomes.* Journal of Family Psychology.
2. Turney, K., & Wildeman, C. (2018). *Incarceration and children's health in the United States.* Social Science & Medicine.
3. Way, N., et al. (2025). *Masculinity norms and barriers to nurturing fatherhood.* Journal of Family Theory & Review.
4. Lee, J., & Hearn, J. (2018). *Caring masculinities: Men, care, and gender equality.* Men and Masculinities.
5. Jones, J., & Mosher, W. (2018). *Fathers' involvement with their children: United States, 2006–2010.* National Center for Health Statistics. https://www.cdc.gov/nchs/data/nhsr/nhsr071.pdf; Centers for Disease Control and Prevention (CDC). (2019). *National Survey of Family Growth, 2015–2019.* https://www.cdc.gov/nchs/nsfg
6. Edin, K., & Nelson, T. J. (2019). *Doing the best I can: Fatherhood in the inner city.* University of California Press.; Roy, K. M., Buckmiller, N., & McDowell, T. (2018). *Together but separate: Partnered fatherhood in low-income Black families.*

Journal of Family Issues, 39(12), 3214–3238. https://doi.org/10.1177/0192513X18763065

7. U.S. Bureau of Labor Statistics. (2023). *Employment status of the civilian population by race, sex, and age.* https://www.bls.gov/cps

8. Connell, R. W., & Messerschmidt, J. W. (2005). *Hegemonic masculinity: Rethinking the concept.* Gender & Society, 19(6), 829–859. https://doi.org/10.1177/0891243205278639

9. Connell, R. W., & Messerschmidt, J. W. (2005). *Hegemonic masculinity: Rethinking the concept.* Gender & Society, 19(6), 829–859. https://doi.org/10.1177/0891243205278639; Griffith, D. M., Gunter, K., & Watkins, D. C. (2018). *Measuring masculinity in health research: A scoping review.* American Journal of Men's Health, 12(2), 414–430. https://doi.org/10.1177/1557988317742739

10. Carson, E. A. (2022). *Prisoners in 2021.* Bureau of Justice Statistics. https://bjs.ojp.gov/library/publications/prisoners-2021

11. Murphey, D., & Cooper, P. M. (2015). Parents behind bars: What happens to their children? Child Trends. Retrieved from https://www.childtrends.org/wp- content/uploads/2015/10/2015- 42Par entsBehindBars.pdf

12. Arditti, J. A., Lambert-Shute, J., & Joest, K. (2019). *Saturday morning at the jail: Implications of incarceration for families and children.* Family Relations, 68(3), 292–306.
https://doi.org/10.1111/fare.12379

13. Elliott, K. (2016). *Caring masculinities: Theorizing an emerging concept.* Men and Masculinities, 19(3), 240–259. https://doi.org/10.1177/1097184X15576203; Heilman, B., Barker, G., & Harrison, A. (2017). *The man box: A study on being a young man in the U.S., UK, and Mexico.* Promundo-US.; Levant, R. F., & Wong, Y. J. (2017). *The psychology of men and masculinities.* APA Books.

14. Releford, Steven A., Stephanie K. Frencher, Antronette K. Yancey, and Keith C. Norris.
"Health Promotion in Barbershops: Balancing Outreach and Research in African American Communities." *Ethnicity & Disease* 20, no. 2 (2010): 185–188.

15. Victor, Ronald G., Katrina Lynch, Ning Li, Christina Blyler, Elijah Muhammad, Jerome Handler, Jessica Brettler, et al. "A Cluster-Randomized Trial of Blood-Pressure Reduction in Black Barbershops." *New England Journal of Medicine* 378, no. 14 (2018): 1291–1301. https://doi.org/10.1056/NEJMoa1717250.

16. Hart, Alvin, and Deborah J. Bowen. "The Feasibility of Partnering with African American Barbershops to Provide Prostate Cancer Education." *Ethnicity & Disease* 14, no. 2 (2004): 269–273.

17. Linnan, Laura A., Heather D'Angelo, and Carolyn B. Harrington.
"A Literature Synthesis of Health Promotion Research in Salons and Barbershops." *American Journal of Preventive Medicine* 47, no. 1 (2014): 77–85. https://doi.org/10.1016/j.amepre.2014.02.007.

18. Alpha Phi Alpha Fraternity, Inc. "National Programs Overview: Go-to-High School, Go-to-College; Project Alpha." Accessed 2023. https://apa1906.net.
19. Rodgers, Autumn B., Carrie P. Morgan, Sharon L. Bronson, Simon Revello, and Tracy L. Bale. 2013. "Paternal Stress Exposure Alters Sperm MicroRNA Content and Reprograms Offspring Stress Axis Regulation." *Journal of Neuroscience* 33 (21): 9003–12.
20. Ciccarelli, Daniela, Valeria Dall'Aglio, Andrea Zanetti, Marco Zoli, and Marco Battaglia. 2020. "Early Life Stress Is Associated with Altered DNA Methylation in Human Sperm." *Translational Psychiatry* 10 (1): 1–12. https://doi.org/10.1038/s41398-020-00887-2.; Laubach, Zachary M., Christopher M. Perng, and Christopher A. Kuzawa. 2019. "Early Life Stress, Paternal Epigenetics, and Intergenerational Health." *Evolution, Medicine, and Public Health* 2019 (1): 1–14. https://doi.org/10.1093/emph/eoz003.
21. Yehuda, Rachel, and Linda M. Bierer. 2009. "Transgenerational Transmission of Cortisol and PTSD Risk." *Progress in Brain Research* 167: 121–35.
22. Kretschmer, Tina, Laura B. Padilla-Walker, and Sarah J. Schoppe-Sullivan. 2023. "Fathers' Stress, Parenting, and Children's Physiological Regulation: A Biopsychosocial Perspective." *Developmental Psychobiology* 65 (2): e22394.
23. Wu, Hsiao-Ying, Chao-Yu Guo, and Wei-Lun Chang. 2022. "Paternal Stress Before Conception Is Associated with DNA Methylation Changes in Human Sperm." *Environmental Epigenetics* 8 (1): dvac006.
24. Margiana, Anggia, Susan D. Calkins, and Esther M. Leerkes. Forthcoming 2025. "Paternal Emotion Regulation and Epigenetic Sensitivity in Early Childhood." *Development and Psychopathology*.
25. Garfield, Craig F., and Anthony Isacco. 2012. "Urban Fathers' Involvement in Their Children's Lives: A Review of the Literature from a Public Health Perspective." *American Journal of Public Health* 102 (6): 100–109.
26. Masarik, April S., Rand D. Conger, Katherine J. Martin, and M. Brent Donnellan. 2021. "Father Involvement and Children's Inflammatory Profiles in Contexts of Stress." *Journal of Family Psychology* 35 (4): 495–507.
27. Sarkadi, Anna, Robert Kristiansson, Frank Oberklaid, and Sven Bremberg. 2008. "Fathers' Involvement and Children's Developmental Outcomes: A Systematic Review of Longitudinal Studies." *Acta Paediatrica* 97 (2): 153–58.
28. Roy, Kevin M. 2006. *Fatherhood in the Margins: Men's Parenting in Low-Income Communities*. New York: Columbia University Press.; Coles, Roberta L. 2009. *The Best Kept Secret: Single Fathers in America*. Thousand Oaks, CA: Sage Publications.
29. Bohacek, Johannes, and Isabelle M. Mansuy. 2015. "Molecular Insights into Transgenerational Non-Genetic Inheritance of Acquired Behaviours." *Nature Reviews Genetics* 16 (11): 641–52. https://doi.org/10.1038/nrg3964.

30. Nestler, Eric J. 2016. "Transgenerational Epigenetic Contributions to Stress Responses: Fact or Fiction?" *PLoS Biology* 14 (3): e1002426. https://doi.org/10.1371/journal.pbio.1002426.
31. Laubach, Zachary M., Brian P. Perera, and Daniel A. Savaiano. 2018. "Transgenerational Epigenetics: Evidence, Mechanisms, and Implications for Human Health." *Evolution, Medicine, and Public Health* 2018 (1): 201–11.https://doi.org/10.1093/emph/eoy018.
32. Bohacek, Johannes, and Isabelle M. Mansuy. 2015. "Molecular Insights into Transgenerational Non-Genetic Inheritance of Acquired Behaviours." *Nature Reviews Genetics* 16 (11): 641–52. https://doi.org/10.1038/nrg3964., Nestler, Eric J. 2016. "Transgenerational Epigenetic Contributions to Stress Responses: Fact or Fiction?" *PLoS Biology* 14 (3): e1002426. https://doi.org/10.1371/journal.pbio.1002426.
33. Sentencing Project. *Parents in Prison: A Snapshot.* Washington, DC: The Sentencing Project, 2023.; Annie E. Casey Foundation. *A Shared Sentence: The Devastating Toll of Parental Incarceration on Kids, Families, and Communities.* Baltimore: Annie E. Casey Foundation, 2024.
34. Bureau of Justice Statistics (BJS). *Prisoners in 2023.* Washington, DC: U.S. Department of Justice, 2023.
35. Sawyer, Wendy, and Peter Wagner. *Mass Incarceration: The Whole Pie 2023.* Northampton, MA: Prison Policy Initiative, 2023.
36. Anderson, Elijah. *Code of the Street: Decency, Violence, and the Moral Life of the Inner City.* New York: W. W. Norton & Company, 1999.
37. Turney, Kristin. "Intergenerational Consequences of Parental Incarceration." *Annual Review of Sociology* 48 (2022): 89–109. https://doi.org/10.1146/annurev-soc-030420-103406
38. Geller, Amanda, Irwin Garfinkel, Carey E. Cooper, and Ronald B. Mincy. "Parental Incarceration and Child Wellbeing: Implications for Urban Families." *Social Service Review* 96, no. 2 (2022): 243–275.
39. Nowotny, Kathryn M., Hedwig Lee, and Christopher Wildeman. "Health and Justice: Framing Incarceration as a Social Determinant of Health." *Annual Review of Criminology* 3 (2020): 313–335. https://doi.org/10.1146/annurev-criminol-011419-041714; Wilper, Andrew P., Steffie Woolhandler, J. Wesley Boyd, Karen E. Lasser, Danny McCormick, David H. Bor, and David U. Himmelstein. "The Health and Health Care of US Prisoners: Results of a Nationwide Survey." *American Journal of Public Health* 113, no. 4 (2023): 458–465.
40. Wang, Emily A., Bruce Western, Donald Berwick, and Marc N. Gourevitch. "Health and Incarceration: The Health of Formerly Incarcerated People in the United States." *New England Journal of Medicine* 383, no. 18 (2020): 1741–1750. https://doi.org/10.1056/NEJMra1911872
41. Binswanger, Ingrid A., Marc F. Stern, Richard A. Deyo, Patrick J. Heagerty, Allen Cheadle, Joann G. Elmore, and Thomas D. Koepsell. "Release from

Prison — A High Risk of Death for Former Inmates." *New England Journal of Medicine* 356, no. 2 (2007): 157–165.; Nowotny, Kathryn M., Hedwig Lee, and Christopher Wildeman. "Health and Justice: Framing Incarceration as a Social Determinant of Health." *Annual Review of Criminology* 3 (2020): 313–335. https://doi.org/10.1146/annurev-criminol-011419-041714

42. Wilper, Andrew P., Steffie Woolhandler, J. Wesley Boyd, Karen E. Lasser, Danny McCormick, David H. Bor, and David U. Himmelstein. "The Health and Health Care of US Prisoners: Results of a Nationwide Survey." *American Journal of Public Health* 113, no. 4 (2023): 458–465.

43. Mitchell, Michael E., and Tonya Davis. "Walking Back into a World That Didn't Wait: Black Fatherhood After Incarceration." *Journal of African American Men* 26, no. 3 (2021): 317–335.

44. Raposa, Elizabeth B., Jean E. Rhodes, Helen J. Stams, and Renee M. Grossman. 2019. "The Effects of Youth Mentoring Programs: A Meta-Analysis of Outcome Studies." *Journal of Youth and Adolescence* 48 (3): 423–443. https://doi.org/10.1007/s10964-018-0962-8.

45. DuBois, David L., and Nils Silverthorn. 2005. "Natural Mentoring Relationships and Adolescent Health: Evidence from a National Study." *American Journal of Public Health* 95 (3): 518–524. https://doi.org/10.2105/AJPH.2003.031476.

46. National Mentoring Resource Center. 2020. *Mentoring Program Effectiveness Review.* U.S. Department of Justice, Office of Juvenile Justice and Delinquency Prevention.
https://nationalmentoringresourcecenter.org.

47. Jarrett, Robin L., J. Luke Miller, and Terri Sullivan. 2011. "Developing Social Capital Through Participation in Organized Activities: Mentoring and the Role of Older Men in Low-Income African American Communities." *Journal of Community Psychology* 39 (3): 305–323. https://doi.org/10.1002/jcop.20430.

48. Griffith, Derek M., Keith A. Thorpe Jr., Keith Allen, and Kristopher Gunter. 2014. "The Influence of Masculinity on African American Men's Health-Seeking Behavior." *American Journal of Men's Health* 9 (1): 1–10. https://doi.org/10.1177/1557988314529540.

49. Natasha J. Cabrera, Catherine S. Tamis-LeMonda, Robert H. Bradley, Sandra Hofferth, and Michael E. Lamb, "Fatherhood in the Twenty-First Century," *Child Development* 71, no. 1 (2000): 127–136; Anna Sarkadi et al., "Fathers' Involvement and Children's Developmental Outcomes: A Systematic Review of Longitudinal Studies," *Acta Paediatrica* 97, no. 2 (2008): 153–158.

Chapter 8: From Streets to Studies—Reclaiming Research, Reclaiming Narratives

1. George, Sheba, et al. 2019. "Structural Racism and Health Inequities." *Health Affairs* 38(9): 1454–1461.; Wallerstein, Nina, et al. 2020. "Shared Power with Communities to Improve Health Equity: CBPR Evidence and Future Directions." *American Journal of Public Health* 110 (S2): S176–S182. https://doi.org/10.2105/AJPH.2020.305818.
2. Griffith, Derek M., et al. 2020. "Men and Health Equity: A Handbook." Routledge.
3. Scharff, Darcell P., et al. 2019. "More than Tuskegee: Understanding Mistrust about Research Participation." *Journal of Health Care for the Poor and Underserved* 30(2): 468–486.; Muhammad, Khalil Gibran, et al. 2021. "Structural Racism and the Criminalization of Health." *The Lancet* 397(10279): 145–146.
4. Wallerstein, Nina, et al. 2018. "Community-Based Participatory Research for Health: Advancing Social and Health Equity." *Annual Review of Public Health* 39: 1–21.; Israel, Barbara A., et al. 2020. "Critical Issues in Developing and Following CBPR Principles." *Health Education & Behavior* 47(1): 24–37.
5. Victor, Ronald G., et al. 2018. "A Cluster-Randomized Trial of Blood-Pressure Reduction in Black Barbershops." *New England Journal of Medicine* 378(14): 1291–1301.; Chatters, Linda M., et al. 2020. "Religion, Stress, and Health Among African American Men." *Journal of Religion and Health* 59(5): 2270–2289.
6. Oetzel, John G., et al. 2022. "Community-Based Participatory Research Outcomes: A Systematic Review." *Health Education Research* 37(1): 1–18.
7. Wallerstein, Nina, et al. 2018. "Community-Based Participatory Research for Health: Advancing Social and Health Equity." *Annual Review of Public Health* 39: 1–21.; Israel, Barbara A., Amy J. Schulz, Edith A. Parker, and Adam B. Becker. 2019. "Review of Community-Based Research: Assessing Partnership Approaches to Improve Public Health." *Annual Review of Public Health* 40: 173–202. https://doi.org/10.1146/annurev-publhealth-040218-044202.
8. Catalani, Caricia, and Meredith Minkler. 2010. "Photovoice: A Review of the Literature in Health and Public Health." *Health Education & Behavior* 37 (3): 424–51. https://doi.org/10.1177/1090198109342084.; Wang, Caroline C., and Mary Ann Burris. 2020. "Empowerment through Photo Novella: Portraits of Participation." *Health Education Quarterly* 24 (2): 171–86.
9. O'Mara-Eves, Alison, et al. 2015. "The Effectiveness of Community Engagement in Public Health Interventions for Disadvantaged Groups: A Meta-Analysis." *BMC Public Health* 15: 129. https://doi.org/10.1186/s12889-015-1352-y.; Muhammad, Mienah Z., Aletha C. Akers, Renee S. Thompson, and Rachel C. Smith. 2021. "Restoring Trustworthiness: Community-Based Participatory Research and the Ethical Obligation to Return Results." *American Journal of Public Health* 111 (4): 657–63. https://doi.org/10.2105/AJPH.2020.306065.

10. Wallerstein, Nina, et al. 2020. *Community-Based Participatory Research for Health: Advancing Social and Health Equity* (3rd ed.). Jossey-Bass.; Wallerstein, Nina, et al. 2020. "Shared Power with Communities to Improve Health Equity: CBPR Evidence and Future Directions." *American Journal of Public Health* 110 (S2): S176–S182. https://doi.org/10.2105/AJPH.2020.305818.; Vaughn, Lisa M., et al. 2018. "Community-Based Participatory Research and Data-to-Action Initiatives." *Progress in Community Health Partnerships* 12 (1): 59–69. https://doi.org/10.1353/cpr.2018.0007.
11. BYP100. 2017. *Agenda to Build Black Futures*. Chicago: Black Youth Project 100.
12. Nina Wallerstein et al., *Community-Based Participatory Research for Health*, 2nd ed. (San Francisco: Jossey-Bass, 2018)
13. Milner, Yeshimabeit. 2019. *Data for Black Lives: A Toolkit for Community Data Justice*. Boston, MA: Data for Black Lives.
14. Milner, Yeshimabeit. 2019. *Data for Black Lives: A Toolkit for Community Data Justice*. Boston, MA: Data for Black Lives.
15. Community Data Lab (CDL). 2022. *Heat Equity Mapping and Participatory Data Projects*. Baltimore, MD: Community Data Lab.
16. Schulz, Amy J., Barbara A. Israel, Edith A. Parker, and Mark A. Lichtenstein. 2013. *Healthy Environments Partnership: A Detroit Community–Academic CBPR Collaboration*.
17. Wallerstein, Nina, et al. 2018. "Community-Based Participatory Research for Health: Advancing Social and Health Equity." *Annual Review of Public Health* 39: 1–21.
18. Muhammad, Mienah Z., Aletha C. Akers, Renee S. Thompson, and Rachel C. Smith. 2021. "Restoring Trustworthiness: Community-Based Participatory Research and the Ethical Obligation to Return Results." *American Journal of Public Health* 111 (4): 657–63. https://doi.org/10.2105/AJPH.2020.306065.
19. Israel, Barbara A., Amy J. Schulz, Edith A. Parker, and Adam B. Becker. 2018. *Methods for Community-Based Participatory Research for Health*. 3rd ed. San Francisco: Jossey-Bass.'
20. Kukutai, Tahu, and John Taylor, eds. 2016. *Indigenous Data Sovereignty: Toward an Agenda*. Canberra: ANU Press. https://doi.org/10.22459/CAEPR38.11.2016
21. Oetzel, John G., et al. 2018. "Participatory Research and Community Outcomes: A Systematic Review." *American Journal of Public Health* 108 (e1): e1–e10.

www.ingramcontent.com/pod-product-compliance
Lightning Source LLC
Chambersburg PA
CBHW070625030426
42337CB00020B/3913